DREAMING IN
CODE

DREAMING IN CODE

Two Dozen Programmers,
Three Years,
4,732 Bugs,
and One Quest
for Transcendent
Software

Scott Rosenberg

THREE RIVERS PRESS • NEW YORK

Three Rivers Press and the Tugboat design are registered trademarks
of Random House, Inc.

Originally published in slightly different form in hardcover in
the United States by Crown Publishers, an imprint of the Crown
Publishing Group, a division of Random House, Inc., New York,
in 2007.

Grateful acknowledgment is made to United Media for permission to
reprint a text excerpt from the comic strip *Dilbert* (May 4, 2004).
Copyright © by Scott Adams. Distributed by United Feature
Syndicate, Inc. Reprinted by permission of United Media.

Library of Congress Cataloging-in-Publication Data
Rosenberg, Scott.
    Dreaming in code : two dozen programmers, three years, 4,732 bugs,
and one quest for transcendent software / Scott Rosenberg.
Includes bibliographical references and index.
1. Computer software—Development.   I. Title.
QA76.76.D47R668 2006
005.1—dc22
2006020614
ISBN 978-1-4000-8247-6

Printed in the United States of America

Design by Maria Elias

10 9 8 7 6 5 4 3 2

First Paperback Edition

For my parents

Software is hard.

—Donald Knuth, author of
The Art of Computer Programming

AUTHOR'S NOTE

The shelves of the world are full of how-to books for software developers. This is not one of them. I'm barely an elementary programmer myself. I wouldn't presume to try to teach the experts. And if my research had uncovered some previously unknown innovation or fail-safe insight into building better software, I'd be smarter to seek investors, not readers.

So while I hope that programmers will enjoy this work, it is meant equally or more for the rest of us. It poses a question and tells a tale. Why is good software so hard to make? Since no one seems to have a definitive answer even now, at the start of the twenty-first century, fifty years deep into the computer era, I offer, by way of exploration, the tale of the making of one piece of software—a story about a group of people setting their shoulders once more to the boulder of code and heaving it up the hill, stymied by obstacles old and new, struggling to make something useful and rich and lasting.

CONTENTS

SOFTWARE TIME

[1975–2000]

I t was winter 1975. I hunched over the teletype in the terminal room, a hulking console that shook each time its typewriter head whammed leftward to begin a new line. I stared at smudgy lines of black code. I'd lost track of the time several hours before; it had to be after midnight. The janitors had switched off the hall lights. I didn't have anyone's permission to hang out there in the NYU physics building, using the computer account that the university was giving away, for free, to high school students. But then nobody was saying no, either.

I was fifteen years old and in love with a game called Sumer, which put me in charge of an ancient city-state in the Fertile Crescent. Today's computer gamers might snicker at its crudity: Its progress consisted of all-capital type pecked out line by line on a paper scroll. You'd make decisions, allocating bushels of grain for feeding and planting, and then the program would tell you how your city was doing year by year. "Hammurabi," it would announce, like an obsequious prime minister who feared beheading, "I beg to report . . ."

Within a couple of days of play I'd exhausted the game's possibilities. But unlike most of the games that captivate teenagers today, Sumer invited tinkering. Anyone could inspect its insides: The game was a set of simple instructions to the computer, stored on a paper tape coded with little rows of eight holes. (The confetti that accumulated in a plastic tray on the teletype's side provided nearly as much fun as the game.) It had somehow landed among my group of friends like a kind of frivolous samizdat, and we shared it without a second thought. Modifying it was almost as easy as playing it if you took the couple of hours required to learn simple Basic: You just loaded the tape's instructions into the computer and started adding lines to the program.

Sumer was a blank canvas—history in the raw, ready to be molded by teenage imaginations. My friends and I took its simple structure and began building additions: Let's have players choose different religions! What if we throw in an occasional bubonic plague? Barbarian invaders would be cool. Hey, what about *catapults*?

That night I was obsessed with rebellion. Sumer had always had a rough provision for popular uprising; if you botched your stewardship badly enough, the people would rise up and remove you from power. (Sumer's original author was an optimist.) I thought there needed to be more variety in the game's insurrections, so I started inventing additions—new subroutines that would plunge Sumer into civil war or introduce rival governments competing for legitimacy.

I didn't care how late it was. The F train ran all night to take me home to Queens. The revolution had to be customized!

■ ■ ■

A quarter century later, in May 2000, I sat in an office in San Francisco and stared at a modern computer screen (high resolution, millions of colors).

Wan ranks of half-guzzled paper coffee cups flanked my keyboard. It was 5:00 A.M.

I was forty years old, a founder and now managing editor of the online magazine Salon, and in charge of a software development project. It had taken us months of meticulous planning. It promised to revolutionize our Web site with dynamic features. And it was disintegrating in front of me.

The lead programmer, after working around the clock for weeks and finally announcing that his work was done, had taken off for Hawaii on a long-planned family vacation. That left his boss, Chad Dickerson, our company's technology VP, to figure out why the database that stored all our Web site's articles would not talk to the programs that were supposed to assemble that data into published pages. Chad had been up for two nights straight trying to fix the problem. Otherwise, our two million readers would find nothing but old news on our site come Monday morning.

Hadn't we built software before? (Yes.) Didn't we test everything? (Not well enough, apparently.) How could we have screwed up so badly? (No good answer.)

I ate the last bag of pretzels from our vending machine, paced, and waited. Nothing helped. There was too much time. Time to read the email from a hapless colleague who'd planned a champagne-and-cake celebration in honor of the new project, and to respond: "Maybe we should hold off." Time to feel alienated and trapped and wonder whether it had been a smart idea to give our system's central server the name Kafka.

We finally published the first edition of our new, "improved" site at around 9:00 A.M. As my coworkers showed up at their desks in their usual Monday morning routines, it took them a while to realize that a half dozen of us had never gone home the night before.

Within a few weeks the software calmed down as our developers fixed the most pressing problems. But every time I hear about a company preparing to "upgrade its platform" by introducing a big new software scheme, I cringe.

■ ■ ■

The 1990s technology-industry boom introduced us to the concept of "Internet time." The phrase meant different things to different people, but mostly it meant *fast*. Under the digital age's new temporal dispensation, everything would happen—technologies would emerge, companies would rise, fortunes would be made—at gasp-inducing speed. That meant you couldn't afford the time to perfect anything—but no need to worry, because nobody else could, either.

Internet time proved fleeting and was quickly displaced by newer coinages untainted by association with a closed-out decade's investment fads. But the buzzword-mongers were on to something. Time really does seem to behave differently around the act of making software. When things go well, you can lose track of passing hours in the state psychologists call "flow." When things go badly, you get stuck, frozen between dimensions, unable to move or see a way forward. Either way, you've left the clock far behind. You're on software time.

■ ■ ■

A novice programmer's first assignment in a new language, typically, is to write a routine known as "Hello World"—a bit of code that successfully conjures the computer's voice and orders it to greet its master by typing those words. In Basic, the simple language of my Sumer game, it looks like this:

```
10 PRINT "HELLO WORLD!"
20 STOP
```

"Hello World" programs are useless but cheerful exercises in ventriloquism; they encourage beginners and speak to the optimist in every programmer. *If I can get it to talk to me, I can get it to do anything!* The Association for Computing Machinery, which is the ABA or AMA of the computer profession, maintains a Web page that lists versions of "Hello World" in nearly two hundred different languages. It's a Rosetta stone for program code.

"Hello World" looks more forbidding in Java, one of the workhorse programming languages in today's business world:

```
class HelloWorld {
    public static void main (String args[]) {
        System.out.println("Hello World!");
    }
}
```

Public static void: gazillions of chunks of program code written in Java include that cryptic sequence. The words carry specific technical meaning. But I've always heard them as a bit of machine poetry, evoking the desolate limbo where software projects that begin with high spirits too often end up.

▪ ▪ ▪

It's difficult not to have a love/hate relationship with computer programming if you have any relationship with it at all. As a teenage gamer, I had tasted the consuming pleasure of coding. As a journalist, I would witness my share of the software world's inexhaustible disaster stories—multinational corporations and government agencies and military-industrial behemoths, all foundering on the iceberg of code. And as a manager, I got to ride my very own desktop *Titanic*.

This discouraging trajectory of twenty-five years of software history may not be a representative experience, but it was mine. Things were supposed to be headed in the opposite direction, according to Silicon Valley's digital utopianism. In the months following our train wreck of a site launch at Salon, that discrepancy began to eat at me.

Programming is no longer in its infancy. We now depend on unfathomably complex software to run our world. Why, after a half century of study and practice, is it still so difficult to produce computer software on time and under budget? To make it reliable and secure? To shape it so that people can learn it easily, and to render it flexible so people can bend it to their needs? Is it just a matter of time and experience? Could some radical breakthrough be right around the corner? Or is there something at the root of what software is, its abstractness and intricateness and malleability, that dooms its makers to a world of intractable delays and ineradicable bugs—some instability or fickleness that will always let us down?

"Software is hard," wrote Donald Knuth, author of the programming field's most respected textbooks. But why?

■ ▮ ▯

Maybe you noticed that I've called this Chapter 0. I did not mean to make an eccentric joke but, rather, to tip my hat to one small difference between computer programmers and the rest of us: Programmers count from zero, not from one. The full explanation for this habit lies in the esoteric realm of the design of the registers inside a computer's central processing unit and the structure of data arrays. But the most forthright explanation I've found comes from a Web page that attempts to explain for the rest of us the ways of the hacker—*hacker* in the word's original sense of "obsessive programming tinkerer" rather than the later, tabloid sense of "digital break-in artist."

Why do programmers count from zero? *Because computers count from zero!* And so programmers train themselves to count that way, too, to avoid a misunderstanding between them and the machines they must instruct. Which is all fine, except for the distressing fact that most of the human beings who use those machines habitually count from one. And so, down in the guts of the system, where data is stored and manipulated—representations of our money and our work lives and our imaginative creations all translated into machine-readable symbols—computer programs and programming languages often include little offsets, translations of "+1" or "–1," to make sure that the list of stuff the computer is counting from zero stays in sync with the list of stuff a human user is counting from one.

In the binary digital world of computers, all information is reduced to sequences of zeros and ones. But there's a space between zero and one, between the way the machine counts and thinks and the way we count and think. When you search for explanations for software's bugs and delays and stubborn resistance to human desires, that space is where you'll find them.

■ ▮ ▮

As this book began taking shape in my thoughts, I would drive to work every day over the San Francisco Bay Bridge. One morning, as my car labored up the lengthy incline that connects the Oakland shore to the elevated center of the bridge's eastern span, I noticed, off to the right, a new object blocking the panorama of blue bay water and green Marin hill: The tip of a high red crane peeked just over the bridge's deck level. It was there the next day and the next, and soon it was joined by a line of a dozen cranes, arrayed along the bridge's north side like a rank of mechanical beasts at the trough, ready to feed on hapless commuters.

Work was beginning on a replacement for this half of the double-decker bridge. A fifty-foot chunk of its upper level had collapsed onto the lower roadway during the 1989 Loma Prieta earthquake. Now a safer, more modern structure would rise next to the old.

In the weeks and months that followed, each of the 240-foot-tall cranes began pounding rust-caked steel tubes, 8 feet in diameter and 300 feet long, into the bay floor. In the early morning hours we could sometimes hear the thuds in my home in the Berkeley hills. One hundred sixty of these enormous piles would be filled with concrete to support the new bridge's viaduct. The whole process was choreographed with precision and executed without a hitch; it felt inevitable, its unfolding infused with all the confidence we place in the word *engineering*.

If the subject of software's flaws is discussed for more than a few minutes at a time, it is a certainty that someone will eventually pound a fist on the table and say, "Why can't we build software the way we build bridges?"

❚ ❚ ❚

Bridges are, with skyscrapers and dams and similar monumental structures, the visible representation of our technical mastery over the physical universe. In the past half century, software has emerged as an invisible yet pervasive counterpart to such world-shaping human artifacts. "Our civilization runs on software," says Bjarne Stroustrup, who invented a widely used computer language called C++.

At first this sounds like an outlandish and self-serving claim. Our civilization got along just fine without Microsoft Windows, right? But software is more than the program you use to send email or compose a report; it has seeped into every cranny of our lives, without many of us noticing. It is in our kitchen gadgets and cars, toys and buildings. Our businesses and banks, our elections and our news media, our movies and our transportation networks, our health care and national defense, our scientific research and basic utility services—the stuff of our daily existence hangs from fragile threads of computer code.

And we pay for their fragility. Software errors cost the U.S. economy about $59.5 billion annually, according to a 2002 study by the National Institute of Standards and Technology, as two out of three projects came in significantly late or over budget or had to be canceled outright.

Our civilization runs on software. Yet the art of creating software continues to be a dark mystery, even to the experts. Never in history have we depended so completely on a product that so few know how to make well. There is a big and sometimes frightening gap between our accelerating dependence on software systems and the steady but slow progress in our knowledge of how to make them soundly. The dependence has increased exponentially, while the skill—and the will to apply it—advances only along a plodding line.

If you talk with programmers about this, prepare for whiplash. On the one hand, you may hear that things have never looked brighter: We have better tools, better testing, better languages, and better methods than ever before! On the other hand, you will also hear that we haven't really made much headway since the dawn of the computer era. In his memoirs, computing pioneer Maurice Wilkes wrote of the moment in 1949 when, hauling punch cards up the stairs to a primitive computer called EDSAC in Cambridge, England, he saw the future: "The realization came over me with full force that a good part of the remainder of my life was going to be spent in finding errors in my own programs." From Wilkes's epiphany to the present, despite a host of innovations, programmers have been stuck with the hard slog of debugging. Their work is one percent inspiration, the rest sweat-drenched detective work; their products are never finished or perfect, just varying degrees of "less broken."

■ ■ ■

Software is a heap of trouble. And yet we can't, and won't, simply power down our computers and walk away. The software that frustrates and hogties us also captivates us with new capabilities and enthralls us with promises of faster, better ways to work and live. There's no going back. We need the stuff more than we hate it.

So we dream of new and better things. The expert who in many ways founded the modern field of software studies, Frederick P. Brooks, Jr., wrote an influential essay in 1987 titled "No Silver Bullet," declaring that, however frustrated we may be with the writing of computer programs, we will never find a magic, transformational breakthrough—we should expect only modest, incremental advances. Brooks's message is hard to argue with but painful to accept, and you can't attend a computer industry conference or browse a programmers' Web site today without bumping into someone who is determined to prove him wrong.

Some dream of ripping down the entire edifice of today's software and replacing it with something new and entirely different. Others simply yearn for programs that will respond less rigidly to the flow of human wishes and actions, for software that does what we want and then gets out of our way, for code that we can count on.

We dream of it, then we try to write it—and all hell breaks loose.

CHAPTER 1

DOOMED

[JULY 2003]

M ichael Toy places his palms on his cheeks, digs his chin into his wrists, squints into his PowerBook, and begins the litany.

"John is doomed. He has five hundred hours of work scheduled between now and the next release. . . . Katie's doomed. She has way more hours than there are in the universe. Brian is majorly doomed. Plus he's only half time. Andy—Andy is the only one who doesn't look doomed. There are no hundreds on his list."

They don't *look* doomed, these programmers sitting around a nondescript conference room table in Belmont, California, on a summer day. They listen quietly to their manager. Toy is a tall man with an impressive gut and a ponytail, but he seems to shrink into a space of dejection as he details how far behind schedule the programmers have fallen. It's July 17, 2003, and he's beginning to feel doomed himself about getting everything done in the less than two months before they are supposed to finish another working version of their project.

"Everybody who has a list with more time than there is in the universe needs to sit down with me and go over it."

These lists are the bug lists—rosters of unsolved or "open" problems or flaws. Together they provide a full accounting of everything these software developers know must be fixed in their product. The bug lists live inside a program called Bugzilla. Toy's programmers are also using Bugzilla to track all the programming tasks that must be finished in order to complete a release of the project; each one is responsible for entering his or her list into Bugzilla along with an estimate of how long each task will take to complete.

"Now let's talk about why we're behind. Does anyone have a story to tell?"

There's silence for a minute. John Anderson, a lanky programming veteran whose title is systems architect and who is, in a de facto sort of way, the project's lead coder, finally speaks up, in a soft voice. "There's a bunch of reasons. In order to build something, you have to have a blueprint. And we don't always have one. Then you hit unexpected problems. It's hard to know how long something's going to take until you know for sure you can build it."

"But you can't just throw up your hands and say, I quit." Toy usually prefers to check things off his agenda fast, running his developers' meetings with a brisk attitude of "let's get out of here as fast as we can" that's popular among programmers. But today he's persistent. He won't let the scheduling problems drop. "We need to make guesses and then figure out what went wrong with our guesses."

Jed Burgess, one of the project's younger programmers, speaks up. "There's a compounding of uncertainty: Your estimates are based on someone else's estimates."

Toy begins reviewing Anderson's bugs. "The famous flicker-free window resizing problem. What's up with that?"

Officially, this was bug number 44 in Bugzilla, originally entered on January 19, 2003, and labeled "Flicker Free window display when resizing windows." I had first heard of the flicker-free window resizing problem at a meeting in February 2003 when the Open Source Applications Foundation

(OSAF), whose programmers Toy was managing, had completed the very earliest version of its project, Chandler—an internal release not for public unveiling that came even before the 0.1 edition. Ultimately, Chandler was supposed to grow up into a powerful "personal information manager" (PIM) for organizing and sharing calendars, email, to-do lists, and all the other stray information in our lives. Right now, the program remained barely embryonic.

At that February meeting, Anderson had briefly mentioned the flicker bug—when you changed the size of a window on the Chandler screen, everything flashed for a second—as a minor matter, something he wanted to investigate and resolve because, though it did not stop the program from working, it offended him aesthetically. Now, nearly six months later, he still hasn't fixed it.

Today Anderson explains that the problem is thornier than he had realized. It isn't simply a matter of fixing code that he or his colleagues have written; its roots lie in a body of software called wxWidgets that the Chandler team has adopted as one of the building blocks of their project. Anderson must either wait for the programmers who run wxWidgets to fix their own code or find a way to work around their flaw.

"So you originally estimated that this would take four hours of work," Toy says. "That seems to have been off by an order of magnitude."

"It's like a treasure hunt," Anderson, unflappable, responds. "You have to find the first thing. You have to get the first clue before you're on your way, and you don't know how long it will take."

"So you originally estimated four hours on this bug. You now have eight hours."

"Sometimes," Anderson offers philosophically, "you just wake up in the morning, an idea pops into your head, and it's done—like that."

Mitchell Kapor has been sitting quietly during the exchange. Kapor is the founder and funder of the Open Source Applications Foundation, and Chandler is his baby. Now he looks up from his black Thinkpad. "Would it be useful to identify issues that have this treasure-hunt aspect? Is there a certain class of task that has this uncertainty?"

"Within the first hour of working on the bug," Burgess volunteers, "you know which it's going to be."

So it is agreed: Bugs that have a black hole–like quality—bugs that you couldn't even begin to say for sure how long they would take to fix—would be tagged in Bugzilla with a special warning label.

Shortly after the meeting, Toy sits down at his desk, calls up the Bugzilla screen, and enters a new keyword for bug number 44, "Flicker Free window display when resizing windows": *scary*.

■ ▪ ▪

Toy's fatalistic language wasn't just a quirk of personality: Gallows humor is a part of programming culture, and he picked up his particular vocabulary during his time at Netscape. Though today Netscape is remembered as the Web browser company whose software and stock touched off the Internet boom, its developers had always viewed themselves as a legion of the doomed, cursed with impossible deadlines and destined to fail.

There was, in truth, nothing especially doomed about OSAF's programmers: Several of them had just returned from a conference where they presented their work to an enthusiastic crowd of their peers—who told them that their vision could be "crisper" but who mostly looked at the blueprint for Chandler and said, "I want it now!" Though the software industry had been slumping for three straight years, they were working for a nonprofit organization funded by $5 million from Kapor. Their project was ambitious, but their ranks included veteran programmers with estimable achievements under their belts. Andy Hertzfeld had written central chunks of the original Macintosh operating system. John Anderson had written one of the first word processors for the Macintosh and later managed the software team at Steve Jobs's Next. Lou Montulli, another Chandler programmer who was not at the meeting, had written key parts of the Netscape browser. They'd all looked doom in the eye before.

Similarly, there was nothing especially scary about bug number 44. It was a routine sort of problem that programmers had accepted responsibility for ever since computer software had migrated from a text-only, one-line-at-a-time universe to today's graphic windows–and-mouse landscape. What scared Toy was not so much the nature of Bug 44 but the impossibility of knowing how long it would take to fix. Take one such unknown, place it next to all the other similar unknowns in Chandler, multiply them by one another, and you have the development manager's nightmare: a "black hole" in the schedule, a time chasm of indeterminate and perhaps unknowable dimensions.

Two months before, the entire Chandler team of a dozen programmers had met for a week of back-to-back meetings to try to solve a set of problems that they had dubbed "snakes"—another word Toy had salvaged from Netscape's ruins. A snake wasn't simply a difficult problem; it was an "important problem that we don't have consensus on how to attack." *Snake* superseded a looser usage at OSAF of the word *dragon* to describe the same phenomenon.

Black holes, snakes, dragons—the metaphors all daubed a layer of mythopoetic heroism over the most mundane of issues: how to schedule multiple programmers so that work actually got done. You could plug numbers into Bugzilla all day long; you could hold one meeting after another about improving the process. Software time remained a snake, and it seemed invincible.

This was hardly news to anyone in the room at OSAF. The peculiar resistance of software projects to routine scheduling is both notorious and widely accepted. In the software development world, lateness was so common that a new euphemism had to be invented for it: *slippage*.

Certainly, every field has its sagas of delay; the snail's pace of lawsuits is legendary, and any building contractor who actually finishes a job on time is met with stares of disbelief. But there's something stranger and more baffling about the way software time bends and twists back on itself like a Möbius strip. Progress seems to move in great spasms and then halt for no reason. You think you're almost done, and then you turn around and six months have passed with no measurable progress.

This is what it feels like: A wire is loose somewhere deep inside the workings. When it's connected, work moves quickly. When it's not, work halts. Everyone on the inside tries painstakingly to figure out which wire is loose, where the outage is, while observers on the outside try to offer helpful suggestions and then, losing their patience, give the whole thing a sharp kick.

Every software project in history has had its loose wires. Every effort to improve the making of software is an effort to keep them tight.

■ ▌ ▌

The earliest and best diagnosis of the problem of software time can be found in a 1975 book by Frederick Brooks, *The Mythical Man-Month*. Brooks was a veteran IBM programming manager who had seen firsthand the follies of the largest software project of its day, the creation of the operating system for the IBM System/360. That huge, expensive mainframe computer would become the mainstay of big business for the next two decades, but its gestation and birth were plagued with delays and cost overruns. It had become, as Brooks put it, a "tar pit," a sticky trap for corporate beasts, even ones as "great and powerful" as IBM. Brooks's group at IBM, confounded by how far behind they'd fallen, flung waves of new programmers at the job, like General Lee's ordering one brigade after another to charge up Cemetery Ridge—only to find that the reinforcements, far from getting the project over its hump, actually made things worse.

And so he formulated Brooks's Law, which is both a principle and a paradox: "Adding manpower to a late software project makes it later."

The soundness of Brooks's Law has been borne out over and over in the three decades since its formulation. But it is so deeply counterintuitive that it still regularly flummoxes programmers and managers, who often would rather pretend that it does not apply to them than deal with its disturbing implications.

Brooks noted that software developers are typically optimists who assume that each bug can be fixed quickly and that the number of new bugs will diminish until the last one has been licked. This optimism, along with the programmers' desire to please the impatient patrons who have commissioned them to create something new, skews the schedule from the start. In practice, Brooks found, nearly all software projects require only one-sixth of their time for the writing of code and fully half their schedule for testing and fixing bugs. But it was a rare project manager who actually planned to allocate developers' time according to such a breakdown.

Next, Brooks argued, the "very unit of effort used in estimating and scheduling" was "a dangerous and deceptive myth." The "man-month" was a concept of scientific management that assumed productivity could be broken down into discrete, identical, fungible units. So if one hundred men (in those days the gender assumption simply came with the territory) could produce fifty widgets in one month, then a single widget required two man-months—and you ought to be able to produce the same number of widgets sooner by throwing more workers at the project. Want your fifty widgets in three days instead of thirty? Just throw one thousand workers at the job instead of one hundred.

Brooks observed that "men and months are interchangeable commodities only when a task can be partitioned among many workers *with no communication among them*." With software, where every project is different and the tools are in constant flux, each time you add a new member to a team, the veterans must drop what they are doing to bring the latecomer up to speed, and everyone needs to pause to reapportion their tasks to give the newcomer something to do. Before you know it, you're even further behind schedule. In the worst cases, Brooks saw, this set up a disastrous loop of delay, a "regenerative scheduling disaster" in which each resetting of the schedule triggers the hiring of more bodies, forcing yet another new schedule into place. Brooks quailed at that prospect: "Therein lies madness."

In any case, Brooks found that much of the work in creating software also suffers from "sequential constraints" that limit how far you can go in splitting up tasks: One task must be completed before the next can be tackled, regardless of how many hands work on it. "The bearing of a child takes nine months," he wrote, "no matter how many women are assigned."

Finally, the extreme variations in productivity from programmer to programmer—a great programmer was typically able to produce ten times as much work in a given amount of time as a mediocre one, and the work was typically five times as good (measured in terms of speed and efficiency)—sealed Brooks's case against the man-month. If one man's work was so drastically different from another's, that yardstick was useless.

■ ■ ■

Brooks's Law implies that the ideal size for a programming team is one—a single developer who never has to stop to communicate with a colleague. This approach streamlines everything, and it also provides insurance that the project will retain what Brooks calls "conceptual integrity": the alignment of all its parts toward the same purpose and according to a harmonious plan. Indeed, the history of software is full of breakthroughs made by lone wolves. But too much software has simply grown too huge for one person to produce—from the elaborateness of the graphic interface that makes programs easier to use to the complexity of the data structures required to reliably and usefully store the huge volumes of information we trust our computers with. Beginning in Brooks's day with operating systems—the basic code that manages a computer system's resources and makes its circuitry responsive to users and their programs—but more recently spreading to every arena of software, programming has become a group effort, a team sport.

The dozen or so programmers at OSAF represented a relatively small team by industry standards, but in their nine months of work since Chandler's official announcement in October 2002, they had already encountered most of the problems that Brooks had outlined decades before. Despite all the computer-based tools at their disposal, from mailing lists and blogs to bug-trackers and source-code version-control systems, staying in sync with one another was fiendishly difficult. Making a schedule was nearly as hard as meeting one. And each new delay introduced the temptation to hire more bodies, yet the new hires never seemed to speed the schedule along.

In one vital respect, however, the work at OSAF represented an entirely new wrinkle on Brooks. The open source software development methodology from which Kapor's foundation took its name simply did not exist in the days of *The Mythical Man-Month*. And more than any other development since then, open source has tantalized the programming world with the prospect of repealing Brooks's Law.

■ ■ ▍

When you buy a typical commercial program—or "shrink-wrapped software"—you receive a file that runs on your computer anytime you need it. This file is a "binary" or "machine-executable" file: It contains machine-level code, impenetrably dense sequences of zeros and ones that is unreadable by most human beings and even most programmers. This binary file is not the direct product of human labor but, rather, the output of a compiler—a computer program that takes lines of code that human programmers wrote in a language like Java or C or C++ and translates them into the machine's language.

Software companies sell a binary file because that's what you need in order to use their product; but they also like it because it protects their secrets. If you want to understand, say, how Microsoft Word was written, you can't find out by peering into its binary file; you would need to see the source code—the thousands of lines of human-written program code that was grist for the compiler. But Microsoft, like most commercial software enterprises, won't let you. The source code to its programs is its most coveted asset, and a Berlin wall of intellectual property law protects the treasure.

That has enabled Bill Gates's company to create the most profitable software franchise in history. It also means that when something goes wrong with Microsoft software, you can't fix it yourself, even if you are an adept programmer. You have to wait for Microsoft to fix it, because only its programmers have access to the code.

The realm of academic computing has always had a more open ethos: Free sharing of source code followed the tradition of free sharing of all scientific research. While Microsoft was coming to dominate the personal-computing software industry through the 1980s and 1990s, this realm left the spotlight, but it never disappeared. When the Internet burst into full bloom on millions of personal-computer desktops in the mid-1990s, it was the university world, with an assist from government funding, that had planted and grown its seeds. In particular, it had developed the protocols that allow easy interconnection between computers made by different companies, built with different hardware, running different operating system software. The software concepts that united computing's isolated archipelagoes into one global network arose not in the offices of profit-seeking entrepreneurs but from the publish-and-share mind-set of idealistic researchers working in universities and publicly sponsored research centers. That world—an environment of geekish enthusiasms and cooperative ideals—experienced a sort of waking to self-consciousness in the 1990s. Programmers looked at one another across the network they had built, blinked, and realized that they shared a set of practices and philosophies different from those of the Microsoft-dominated personal-computing mainstream. They took their inspiration and tools from two central figures: Richard Stallman and Linus Torvalds.

In 1985, Stallman, an eccentric MIT genius who was irate about the commercial software industry's habit of locking up code, established the Free Software Foundation. It developed a special kind of software license that said you could have all the code you wanted, and reuse it, and incorporate it into new products—but anything you created with that code had to be covered, in turn, by the same licensing terms. This license, called the GPL (for GNU Public License), was explicitly designed to limit the privatization of free programs, and its critics came to view it as a sort of nasty infectious virus; Bill Gates derided it as anticapitalist and compared it to Pac-Man, the video game critter with the vast appetite. But for Stallman and the programmers who joined his band, it was an effective barricade against the encroachments of commerce.

Stallman and his cohorts labored for years in relative obscurity trying to build the pieces of GNU—a project to create a free version of the Unix operating system that predominated in university computing centers. Meanwhile, a young Finnish programmer named Linus Torvalds began posting to Internet newsgroups about a little operating-system project of his own—"just a hobby, won't be big and professional like gnu," he declared in a 1991 message. As the project attracted volunteer contributors, Torvalds welcomed them and gradually shaped them into what is now known as the Linux community. Linux provided GNU with the central component it was missing, the operating system's "kernel," the focal core of the digital brain.

In the 1990s, the operating system that purists call "GNU-Linux" steadily matured into a stable, free version of Unix, with its innards laid bare. And as the business world began to embrace it, the programmers who built it started examining their own methods, formalizing their practices and giving them a name: open source software development.

Proponents of open source like to draw the distinction between "free as in free beer" and "free as in free speech": Not all open source software products cost nothing, but all open source software is free to be examined, adapted, and reused. The open source world is full of contentious camps: Stallman's followers, for instance, view the phrase "open source" as a sellout and prefer the "free software" label; and debates over differing licensing terms are pursued with religious fervor. But advocates of every stripe share one article of faith: that software anyone can tinker with is bound to improve over time in ways that "closed" software can't match.

■ ■ ■

Open source is widely portrayed in the media as a challenge to Microsoft, and it is. But the challenge is indirect. Commercial software is software that is designed to turn a profit, and it is hard to imagine anyone figuring out how to do that better than Microsoft. The only way to take Microsoft on is to change the game's objective. What if the point of software wasn't profit; what if you just gave programmers a chance to have fun and show their chops? Of course this sounds absurdly utopian. But what if self-motivated programmers actually produced better code? And what if they did so more efficiently?

Open source doesn't just offer an alternative economic basis for producing and distributing software; it can radically change the nuts-and-bolts process of developing software—moving it from the cloistered few to a distributed crowd. "The Cathedral and the Bazaar," a 1997 essay by programmer Eric S. Raymond, remains the most cogent explanation of that change. Raymond, a longtime hacker of the old school, described the epiphany he felt when he first realized the implications of Torvalds's techniques:

> I had been preaching the Unix gospel of small tools, rapid prototyping and evolutionary programming for years. But I also believed there was a certain critical complexity above which a more centralized, a priori approach was required. I believed that the most important software . . . needed to be built like cathedrals, carefully crafted by individual wizards or small bands of mages working in splendid isolation, with no beta to be released before its time. Linus Torvalds's style of development—release early and often, delegate everything you can, be open to the point of promiscuity—came as a surprise. No quiet, reverent cathedral-building here—rather, the Linux community seemed to resemble a great babbling bazaar of differing agendas and approaches . . . out of which a coherent and stable system could seemingly emerge only by a succession of miracles. The fact that this bazaar style seemed to work, and work well, came as a distinct shock.

"The Cathedral and the Bazaar" dissected Torvalds's style of managing the Linux project over (then) half a decade's development from personal hobby to global phenomenon and derived a set of principles from it. "Every good work of software starts by scratching a developer's personal itch." Programmers are motivated and led toward their best work by a desire to accomplish something that pleases them or fulfills a personal need. "Good programmers know what to write. Great ones know what to rewrite (and reuse)." No programmer will want to build something from scratch if he can grab something that is already written and adapt it to new ends.

In the Unix world, programmers had long been accustomed to freely sharing their source code. But for important work, cathedral mode had long remained the default, for both free and commercial software. The most important predecessor to Linux, BSD Unix, was largely the work of one man, Bill Joy, who went on to cofound Sun Microsystems. And Stallman's Free Software Foundation produced its popular GNU software tools—like the emacs editor found on so many programmers' desktops to this day—under a cathedral approach, too. Torvalds and Linux changed that, demonstrating that a "promiscuous," bazaar-style approach could produce big, useful software that kept getting better. By the time Raymond wrote "The Cathedral and the Bazaar," the successes of Linux and other open source software projects—such as the Apache Web server (with, at the time I write this, a roughly two-thirds market share compared to 25 percent for Microsoft's competing product)—were already beginning to make heads turn.

But Raymond was the first observer to see, in Torvalds's work, an explicit transcendence of Brooks's Law—a new mode of working on software that allowed you to harness the skills of a large number of programmers without dragging your project into the mire. Raymond identified two key prerequisites for this: cheap, widespread access to a network, like the Internet, that enabled fast, reliable communication among developers and storage of common, openly accessible pools of shared knowledge and code; and the rise of a cooperative group ethos built around a leadership style, like Torvalds's, that encouraged newcomers, welcomed contributions, and strove to maximize the number of qualified participants.

Once you have established these conditions, Raymond wrote, you could drastically improve the process of finding and fixing bugs in any program. Instead of jealously hoarding your code out of fear of competitive poaching, you could hang it out for the whole world and invite your colleagues in—and if you had something compelling and knew how to work the developer crowd, a self-reinforcing feedback loop of collective expertise would kick in.

"Given a large enough beta-tester and codeveloper base, almost every problem will be characterized quickly and the fix obvious to someone. Or, less formally, 'Given enough eyeballs, all bugs are shallow.' I dub this: 'Linus's Law.'" Raymond argued that this new style of network-powered open peer review had broken the back of Frederick Brooks's cruel paradox. "To Brooks's Law I counterpropose the following: Provided the development coordinator has a communications medium at least as good as the Internet, and knows how to lead without coercion, many heads are inevitably better than one."

Raymond's essay caught and then fueled a wave of excitement in the programming world. The IBM mainframe projects that Brooks cut—or broke—his teeth on precisely fit Raymond's description of a "cathedral style" of programming. In the late 1990s, cathedrals were still being built, especially in Redmond, Washington, by Microsoft's priesthood. But the hubbub from the tent cities beyond their gates was getting hard to ignore.

Still, the passage from cathedral-style logjam to the efficiency of open source bazaars was not quite as simple or as painless as open source's most fervid advocates would have it. For one thing, as Raymond acknowledged, Brooks himself had foreshadowed at least some of the principles that drove the new movement: In *The Mythical Man-Month*, he had written, "The total cost of maintaining a widely used program is typically 40 percent or more of the cost of developing it. Surprisingly this cost is strongly affected by the number of users. More users find more bugs." In fact, in the programming world of Brooks's youth, free availability of a program's source code was the norm: Software was not then the basis of a multibillion-dollar industry; it was simply something you obtained from a computer's manufacturer so that the computer would be usable and useful, and it was freely shared. In his 1975 book, completed right before the birth of the personal-computer software industry and the founding of Microsoft, Brooks defined a full-fledged "programming product" as one well enough documented that "anyone may use it, fix it, and extend it"—which sounds more like the work of Linus Torvalds than of Bill Gates. Brooks also complained about programs that are "incompletely delivered (no source code or test cases)." A complete program, to him, was one that provided its source as a basis for further work.

So open source itself, the availability of source code for programmers to study and extend, was already a part of the world Brooks chronicled. But the open source methodology and the Internet-based collaborative bazaar still lay in the future. Raymond's key insight was less a matter of explaining why it was useful for programmers to have access to source code than of identifying the importance of the Internet, and of Torvalds-style leadership, in making that access to code valuable.

"The Cathedral and the Bazaar" made a persuasive case for Torvalds's brand of open source as a leap forward, but it didn't fully come to grips with the difficulties Brooks had identified in the process of building new software from scratch. Raymond showed how the open bazaar could harness the talents of a larger number of programmers without activating the harsh logic of Brooks's Law, but he couldn't show that open source made it any easier to predict how much time it would take to write a new program or to speed its delivery to a waiting public. "The Cathedral and the Bazaar" doesn't actually repeal Brooks's Law and solve the problem of time in software development; it maps an alternate universe for programming in which time is simply less important because the work is cooperative rather than corporate, the workers are all volunteers, and the motivation is fun and ego, not financial reward.

"Enjoyment predicts efficiency," Raymond wrote. "Joy is an asset." The idealistic maxims are inspiring, but they have yet to be conclusively proven in the field. Open source seems to suffer from its own species of frustrating delays. The highest profile new open source project in the years since the publication of Raymond's essay was Mozilla—a Web browser built out of Netscape's corporate wreckage. In the last throes of its battle with Microsoft, Netscape had embraced open source and published its browser's source code. The result today is a still-evolving free program used by millions. Its most recent incarnation, Firefox, has actually begun eating into the market share of Microsoft's Internet Explorer. But Mozilla took an agonizingly long time to become that valuable, and meanwhile the competition at Microsoft cemented its dominance. The new bazaars of the open source movement have changed computing in many ways, but they are not notable for bringing new products to users any faster than the old cathedral builders did.

And bringing a new product to users was what Kapor and his Chandler developers ached to do in the summer of 2003. They were an odd hybrid: They functioned like an open source project in that they were posting their source code on the Internet and trying to build a community of volunteer developers around it; but they also felt and acted like a classic software start-up company, with a core group of programmers trying to get a new product off the ground and worrying about how long it was taking. For Kapor, who had built Lotus Development Corporation from the ground up and later spent a couple of years as a venture capitalist, the start-up model was the world he knew, and open source was the world he was groping toward.

His motivation to create Chandler—an ambitious rethinking of PIM software to enable easy sharing of data and to run on the three most popular kinds of personal computers—was, undoubtedly, to use Raymond's phrase, "scratching a personal itch." He wanted to build Chandler because he craved something like it, and similar existing products simply didn't match the picture in his head. But he didn't begin, as a lone open source hacker might have, by posting a first chunk of code; he began, as an entrepreneur would, by subleasing some office space.

As with so many software start-ups, OSAF's first digs were funky, informal, improvised. Kapor had found space in Belmont on the Peninsula fringe of Silicon Valley, in a nondescript low-slung office park just down the road from the glass-towered campus of Oracle, the giant producer of corporate databases. The building also housed "Oracle University." An Extended Stay America motel was going up next door for migrant businesspeople stopping off at Oracle. But the office complex had plenty of vacancies in the spring of 2002 when Kapor first moved his fledgling company into a corner of the fourth floor, using space no longer needed by a software firm called Reactivity. That's the way Silicon Valley replenishes itself, forest-like: The technology industry moves in a perennial cycle of growth and decomposition, and new companies have always germinated in the empty space created by the toppling of other companies from a previous wave of growth.

OSAF's developers spread out in the corporate void, filling room-length whiteboards with their scrawls. When they ran out of space, they applied their dry markers to the tinted windows. At the time I first visited them in January 2003, they had the appearance of homesteaders on an office prairie, huddled around the cool campfires of their monitors, trying to spark a revolution.

In July 2003, OSAF moved out of its Belmont space into offices in downtown San Francisco, on the top floor of an old loft building that had been completely remodeled according to eco-friendly principles: recycled wood products and solar heating. It also had a dog run on the roof for the "dog-friendly environment" that Kapor sought. Many of the developers owned dogs, and the previous year he and his wife had acquired a pair of frisky labradoodles named Cosmo and Chandler. (The hound took his name from the program, not vice versa.)

The new space was beautiful, with big exposed beams in the ceilings, but also—as new, unlived-in offices often are—a bit forlorn. Someone had written in huge letters across a whiteboard: "Welcome to Howard Street." Later, someone else erased a bunch of letters and added one so that the board now read: "We come to ward three." That was too bleak for another staffer, who erased an *r*, leaving a more positive biblical prophecy: "We come toward thee."

■ ■ ■

Kapor repeatedly told his programmers that the nonprofit OSAF was going to operate under different rules from the venture-capital-funded start-ups whose wreckage, in those post–Internet boom days, littered the San Francisco Bay area—leaving OSAF's new neighborhood south of Market Street feeling like an urban dead zone. At a May meeting he reassured some of them who feared that their team was growing too big, too fast: "We're not operating with the mythology of the Silicon Valley death march, with the deadline to ship a product and get revenue, where the product quality goes out the window. Everybody who's been through that—and that pretty much includes everyone in this room—knows what that's like. If we go down that route, we will have failed miserably."

Kapor kept this promise: OSAF's growth was steady but careful, and indeed by summer's end many programmers worried that hiring was too slow. In theory, the open source model meant that volunteers would throng to push work on Chandler forward, and a handful of programmers showed up regularly on OSAF's open Internet Relay Chat channel and paid attention to its interim code releases. But there wasn't much that such outside contributors could do yet. "We're just playing at open source," I'd hear from more than one of the Chandler crew during the slow months of 2003, when it sometimes seemed as though the project was becalmed. Chandler—neither driven by investor-dictated deadlines nor fueled by a critical mass of open source code—was caught in a dispiriting limbo.

In June, Michael Toy started a blog musing on his work as Chandler's development manager. In his first posting, he pondered why progress on Chandler seemed so agonizingly slow. Contrasting OSAF with his experience at Netscape, he laid part of the blame on OSAF's more democratic, less hierarchical structure—the very structure that "The Cathedral and the Bazaar" had championed as an efficient technique for marshaling stubborn programmers: "At OSAF we rule by general consensus, and decisions happen slowly. The most important thing is that we make good decisions. At Netscape the most important thing was that we needed to write code, and so almost any decision which unblocked people and let code get written was the right one. The problem at OSAF is that it seems really hard to clear a path so that code can be written."

Part of the difference, Toy felt, lay in Chandler's novelty. It was not a rewrite like so many open source projects; it had grand ambitions. "At Netscape the browser was a known quantity, it just needed to be re-written. At OSAF, much of what will make Chandler special is a new and unique approach to data. The uncertainty about exactly what this means seems to pop up as part of every decision, the ground on which we are trying to stand is still cooling."

Thirty years before, Frederick Brooks had observed, "The hard thing about building software is deciding what to say, not saying it." OSAF was proving the maxim's continued truth.

I sat down with Andy Hertzfeld six weeks after Toy had vented his frustrations in his blog. At OSAF meetings Hertzfeld's voice was the most consistent in pushing the developers to stop designing and start coding—or at least to start coding without waiting for the ground to cool. "My style," Hertzfeld told me, "is to get something going really quick and then turn it into the great thing that is the reason you're doing it. You're not working on it to have it be run-of-the-mill. You're working on it to do something great. But you need to get it started! The key is getting exciting work going; the rest of it will take care of itself. You're sparking off each other—a virtuous cycle—once you're doing the thing you're there to do."

Toy had framed the question: "Is it possible to do great work without great pressure, or is pressure an indispensable part of genius?"

■ ■ ■

In between his musings about Chandler, Toy sometimes used his blog as an outlet for his enthusiasm for baseball, even posting photos from a Giants game that he had slipped off to one Thursday afternoon in August. If he'd had time to take a copy of *The Mythical Man-Month* with him to Pac Bell Park (as it was then known) and flipped through it between innings, he would have found the following passage: "A baseball manager recognizes a nonphysical talent, *hustle*, as an essential gift of great players and great teams. It is the characteristic of running faster than necessary, moving sooner than necessary, trying harder than necessary. It is essential for great programming teams, too."

With nine months of work under their belt and a product that remained more vision than code, the Chandler programmers still hadn't found their hustle.

"Joy is an asset," Eric Raymond had declared. "It may well turn out that one of the most important effects of open source's success will be to teach us that play is the most economically efficient mode of creative work."

Stirring words. But as summer drew to a close, there was, as yet, no joy in Chandlerville.

CHAPTER 2

THE SOUL OF AGENDA

[1968–2001]

One Thursday morning, a dozen or so members of the Chandler development team gathered for the weekly staff meeting in the big conference room on the fifth floor of 543 Howard Street in San Francisco. Laptops opened up and locked on to WiFi signals, coffee cups steamed, dogs roamed, and people chatted. Then, with no fanfare, Mitch Kapor poked his head through the double doors and said, "Here's someone you know." He led in an oddly familiar-looking tall man in a blue suit who stood stiffly for a moment while the chatter subsided.

Then the visitor spoke. "I'm Al Gore. I used to be the next president of the United States." The only jaws that didn't drop were the dogs'.

The former vice president had shown up to attend a meeting Kapor was hosting for a socially responsible investment fund. He seemed a little unsure why he was now standing in front of a bunch of programmers, and he didn't have a lot to say other than "Keep up the great work! You guys are changing the world!" Then he stepped out again.

It wasn't clear how much Gore actually knew about Chandler and OSAF, but guessing that their goal was "changing the world" was a pretty good bet. In Silicon Valley that phrase had been a rallying cry at least since Steve Jobs's legendary pitch to Pepsi CEO John Sculley twenty years before, asking the exec whether he preferred to "sell sugar water" the rest of his life or come build computers and change the world.

To a jaded ear it might sound impossibly naive, but the people working on Chandler—like, in my experience, software developers everywhere—were motivated by the hope that their work might make a difference in people's lives. Perhaps the idealism most programmers share is a direct consequence of the toil and frustration of programming. If you're going to have to wrestle with daunting abstractions or squash armies of bugs, big ambitions can help pull you through the slog.

Although Gore might have gotten the tense wrong—Chandler wasn't actually changing the world (not yet, anyway)—he got the aspiration right. Chandler was indeed fueled by world-changing dreams. But it began more modestly, with a small irritation, an everyday annoyance—an itch.

■ ■ ■

Back in 2001, a Microsoft Exchange server sat in a closet just outside Mitch Kapor's office in downtown San Francisco. The closet was hardly where Kapor wanted to spend his time, but as the first responder when there were problems with Exchange, that was where he frequently found himself at the start of the new millennium, often at odd hours. Kapor's wife, Freada Kapor Klein, ran a small nonprofit organization from the same office suite, and staff members used Microsoft's software to share calendars as well as to manage email.

Exchange is a big program capable of handling the flow of data for companies hundreds of times this size; it is the beating heart of most "Microsoft shops"—firms that have built their software universes around

Bill Gates's products. Kapor's antipathy to Gates and everything the Microsoft founder stood for in the software field was a matter of record. He was not happy about using Exchange, but at the time—in 2000 and 2001, as the Silicon Valley tech boom peaked and then nosedived—it was the least unbearable of available options for small-group scheduling. Still, it seemed crazy, far more powerful than the office needed. Costlier, too: You had to dedicate a computer as the server, you had to pay for a Windows license for that server, you had to license the Exchange software itself, and if you didn't have a full-time technician on staff, you had to pay a consultant to come in for a few hours every month and give the thing a tune-up. Before you knew it, you were spending thousands of dollars just to keep a handful of calendars in sync.

There had to be a better way. Why did you have to invite Microsoft to the party? Why couldn't the calendar on your computer talk directly to the one on your coworker's machine? In that sort of "peer-to-peer" approach, individual computers form a network by connecting directly to one another. The server was just a middleman; middlemen could be cut out.

Peer-to-peer had recently become a hot notion in the online world thanks to the overnight success of Napster, which introduced a limited version of the concept to millions of people by showing them how easily they could trade music files. The music industry and the courts stepped in and shut Napster down, but by then the peer-to-peer vision had taken hold. For one thing, it worked. For another, it scaled: You could apply the principle to ever higher numbers of computers and users, and it still worked, without anyone's having to spend a fortune on some central facility. Finally, it held the same antiauthoritarian appeal that had fueled the early days of the personal computer revolution: a promise to dethrone the high priests of computing, the gurus who had run the big mainframe computers decades before and now guarded the door of the server room, and empower Everyman to do amazing things with the box on his desk.

That, after all, was where Kapor's own trajectory as a personal-computing pioneer had started two decades before. Now he knew he wanted to make software again, and he'd been wondering what to tackle. Maybe this was it.

■ ■ ■

To anyone who followed the explosive rise of personal computing in the early 1980s, when *Time* magazine let the PC elbow human competitors off its year-end cover to become 1982's "Machine of the Year," Kapor's name will be forever associated with a program called Lotus 1-2-3. It wasn't the first spreadsheet; that milestone belonged to Dan Bricklin's VisiCalc, which showed the world of business that microcomputers, small desktop machines with less memory and power than the previous generation of refrigerator-sized minicomputers, could do useful work. What Lotus 1-2-3 offered the world was unusual flexibility and legendarily good design. VisiCalc ran mostly on the popular Apple II computer, but 1-2-3 was built specifically for the new IBM PC. IBM's first entrant in the desktop computer market was still a rarity in most offices, but Kapor bet his business on it and built 1-2-3 to show off its power and speed. The program proved so popular that people began to buy new IBM PCs just so they could run Lotus's software. It was the embodiment of the computer industry cliché of the "killer app."

"Killer," however, was not the kind of term the company's founder had in mind when he chose its Buddhist-inspired name. Lotus 1-2-3 transformed the company into an overnight success—for a time in the eighties it was bigger than Microsoft—and made Kapor rich ($100 million or so at the time, according to *Business Week*). But it didn't make him happy. As he explained to interviewer David Gans in 1990, when the Lotus saga was still a fresh memory: "It's important to understand that before I was a digital capitalist I used to teach meditation, and then I was a counselor in the psych unit of a local community hospital, which was a formative experience. I have a master's degree in counseling psychology. So I've been pretty much all over the map. I just kind of fell into computers; I didn't set out to be Bill Gates—Bill Gates set out to be Bill Gates. My perspective was really never totally shaped by needing to succeed in building a big company and making a lot of money. In a nutshell, I started this little company called Lotus and made this software product that several million people wound up buying, and this little company turned into this enormous thing with thousands of employees making hundreds of millions of dollars a year. And it was awful. It felt awful to me personally. So I left. I just walked away one day."

Kapor resigned from Lotus in July 1986—"extricating myself from my own success," as he described it to *Inc.* magazine at the time. But before he walked away, there was a project he still cared about, and he stuck around Lotus just long enough to finish it. The idea was as ambitious as computer-based artificial intelligence and as mundane as the pile of small papers—business cards, Post-its, notebook pages—that accumulated every day in the breast pocket of Kapor's Hawaiian shirts. Not all of us are buried in scraps of paper, but life throws a barrage of details at everyone. Computers are better at remembering small details than human beings. Couldn't PCs help us stay organized? You could buy calendar programs and address books and the like, but what about all the random, hard-to-categorize information that flows into a busy person's life each day?

The Lotus project, called Agenda, aimed to solve Kapor's little-pieces-of-paper problem. Computers tend to store information rigidly, in set, predefined categories, a system that works reasonably well for business inventories, financial accounts, and phone directories. The papers in Kapor's pocket—reminders, things to do, ideas, recommendations, and so on—weren't so easy to categorize. He wanted to be able to dump them into the computer on the fly and worry about organizing them later. And he needed to count on being able to find them easily and quickly.

Lotus Agenda, the program that was officially unveiled in 1988, was an elegantly simple list maker with a few unusual features that elevated it into a software legend. You could use it to manage your daily life, but you could also use it to organize academic research, a music collection, a work project, anything that involved many small bits of information. Its "automatic assignment" feature—given a vague phrase like "lunch with John next Friday," it could figure out the date of "next Friday"—felt like magic; it's still hard to find software that matches it. And it introduced a unique new approach to organizing data—halfway between the rigid structures of traditional computer databases and the free format of word processing.

Agenda broke new ground in the no-man's-land that separated the strict realm of computer logic and the fuzzy ambiguities of human reality. The principles that drove its creators (Kapor worked with Jerry Kaplan, who went on to found the pen-computing start-up Go, and Edward J. Belove) were ambitious: Users should be able to input their data without worrying about the structure of the software that will store it; users should be able to extend and modify the structure of the data easily, adding new categories without losing any information; users should be able to build new ways of viewing their data and to manipulate and change the data through those views that they created themselves.

You can gauge just how bold these goals were by seeing how little of the software that we now employ, two decades later, meets them. Every time a program or a Web site demands that you fill in a blank field *its* way rather than *your* way—don't put a hyphen in that Social Security number! don't include the spaces in that credit card number!—you are encountering a problem that Agenda, in its own idiosyncratic way, had solved.

Agenda won rave reviews and developed a cult following. Paeans to it still dot the Web, including one from a 1992 article in *The Atlantic* by James

Fallows that reads: "Of all the computer programs I have tried, Agenda is far and away the most interesting, and is one of the two or three most valuable." For all the devotion it inspired, though, Agenda never achieved "killer" status. This was in part because Lotus, which had become a corporate behemoth looking for "big wins" in the business market, didn't quite know what to do with it and was perhaps not enthusiastic about selling the brainchild of a founder who had forsaken the company. Also, Agenda ran under DOS, Microsoft's crude original operating system for the PC, and the future plainly belonged to the graphical world of Macintosh- and Windows-style systems. (Kapor says Lotus developed a Windows version of Agenda but never released it.) Lotus's marketing department invented the term "Personal Information Manager" to describe the new software category Agenda had opened up, and the term is with us today, in acronym form: Microsoft Outlook is still considered a PIM. But as the 1990s wore on, Lotus "orphaned" Agenda, as the software industry says of programs that are no longer being improved or supported by their makers.

Maybe the program was ahead of its time. Maybe it was an unnecessary victim of misplaced corporate priorities. Or perhaps, for all its innovation and flexibility, it was a little *too* versatile for most users—as Fallows put it, "too powerful for mortal men."

■ ▨ ▨

Kapor moved on from Lotus. He taught software design briefly at MIT. He founded a new venture called On Technology that developed software for people to collaborate across a computer network. He was a key founder (and funder) of the Electronic Frontier Foundation, the online civil liberties lobby. He made some smart, farsighted investments in the early Internet era in companies that would ride the wave of the nineties boom, and he campaigned for an open network at a time when most experts were still embracing the "walled garden" approach of closed-network operations like America Online, CompuServe, and Prodigy. Later, he moved from Boston to Silicon Valley and became a venture capitalist at Accel Partners, a role he quickly found himself unsuited for. ("My heart is with the entrepreneur," he explained to me later, "and when you're a VC, there are too many times that you have to act in a contrary way. You're serving the interests of the limited partners [chief investors], and sometimes the interests are aligned, but sometimes they're not.")

But he never really moved on from Agenda. He remained, he'd say, "so emotionally attached to this product, it's like my child." As he began winding down his career as a venture capitalist, he contracted with some programmers to build an up-to-date version of Agenda using Java. But the result was unsatisfying. "It felt narrow, old-fashioned," Kapor says. "It was kind of like building a model-T Ford. We know a lot more about cars now!" It turned out that what he cherished was not the "feature set" of Agenda— the list of specific things it could do—but the program's spirit of dynamic flexibility, of "put it in first, make decisions later." Whatever shape his new software would take, Kapor decided, as he pondered the inadequacies of Microsoft Exchange and began to dream of inventing something to put in its place, it would have to conjure the soul of Agenda.

The only other thing he was sure of at that point, as the seeds of the project that would become Chandler began to sprout, was that he wanted the new product to be open source. In the spring of 2001, the Internet's new economy lay in ruins, and old-fashioned, for-profit start-up software companies were out of market favor. No one at the time thought there was a future in challenging the dominance of Microsoft's Exchange and Outlook programs in the world of email and personal information management. But Kapor's attraction to open source wasn't simple opportunism.

He had followed the rise of the open source movement in the late nineties; he found its ideas congenial, and even more important, he found its arguments persuasive.

This was not an obvious or inevitable choice for someone whose career epitomized the triumph of the entrepreneurial software capitalist and whose fortune originated in the sale of shrink-wrapped programs. Back in the 1980s, in his Lotus days, Kapor had sat miserably in his office while Richard Stallman, the disheveled and cantankerous torchbearer of the free software movement, led a crowd of chanting picketers from the League for Programming Freedom in the street outside. They were protesting Lotus's policies on software copyrights and "look-and-feel lawsuits" that tried to block other programmers from mimicking user interface features from Lotus programs. Kapor says he actually sympathized with the protest. But even though he was no longer the CEO, he felt obligated to support the company he had founded; it was just this sort of "intolerable personal conflict" that led him to leave the company for good.

Now there were no such conflicts. Kapor had become a careful philanthropist, and the open source approach gave him the chance to feel he was making a public contribution while doing the kind of software design that he loved. "I had lunch with Linus Torvalds and other people," Kapor recalls, "and I convinced myself that it wasn't crazy." The very improbability of the open source movement—you mean nobody owns the code? and you just give it away?—appealed to his maverick idealism. And Torvalds's results-oriented case for open source appealed to his businessman's pragmatism.

The open source movement had already been through several rounds of a complicated dance with the business world. At the height of the Internet stock market craze in the spring of 2000, Linux had experienced its own mini-bubble, and a number of open source companies, including Red Hat and VA Linux, had gone public. Open source stocks fared no better than any other technology investments during the dot-com wipeout that followed. But the corporations of the world still needed to run computer systems, and in the era of belt tightening that started with the market crash of 2000 and deepened in the days after 9/11, the price tag of open source-based systems looked very attractive.

In beginning to embrace Linux and its ilk, businesses had to contend with several arguments against open source. Beware, said critics: Without a company behind the code, you'll never get help when you need it. In fact, there were more and more companies offering service contracts on open source systems, and more and more programmers and system administrators who knew them inside and out. The critics also decried the tendency of open source projects to "fork"—to respond to technical disagreements by splitting into rival camps. Since no one owned the code, anyone could take it in a new direction. This sometimes led to confusing proliferations of similar sounding products; but as long as there was a core project that held the allegiance of skilled developers who continued to improve the code, why should that worry anyone? Finally, critics argued, open source methods still hadn't proved their value in developing products that were usable for people who didn't live and breathe computer technology.

That complaint, because it was largely true, stung the most. And Kapor hoped his new project would answer it.

Torvalds, who is known as Benevolent Dictator for Life of the Linux operating system, consistently exudes a calm optimism about the long-term prospects for the movement he symbolizes. "In science," as he explained in a 2004 interview in *Business Week*, "the whole system builds on people looking at other people's results and building on top of them. In witchcraft, somebody had a small secret and guarded it—but never allowed others to really understand it and build on it. Traditional software is like witchcraft. In history, witchcraft just died out. The same will happen in software. When problems get serious enough, you can't have one person or one company guarding their secrets. You have to have everybody share in knowledge."

■ ■ ■

In the spring of 2001, Mitch Kapor established the Open Source Applications Foundation as a nonprofit organization and hired its first employees. He needed, as he later put it, some "people with good pairs of hands" to begin exploring ideas. He hired two young programmers, Morgen Sagen and Al Cho, who had worked together at the dot-com flameout Excite@Home.

At this point Kapor knew only three things about what he wanted to build: It would be open source. It would scratch the Exchange itch. And it would carry the soul of Agenda into the future. Beyond that, it was a blank slate.

■ ■ ■

The dream of using the computer as a tool to master tides of information is as old as computing itself. In a 1945 essay titled "As We May Think," Vannevar Bush, who oversaw the U.S. government's World War II research program, unveiled his blueprint for the Memex, a desk console with tape recorders in its guts that would give a researcher ready access to a personal trove of knowledge. Bush's Memex provided the nascent field of computing with its very own grail. For decades it would inspire visionary inventors to devise balky new technologies in an effort to deliver an upgrade to the human brain.

By far the most ambitious and influential acolyte of the Memex dream was Douglas Engelbart, best known today as the father of the computer mouse. Engelbart, a former radar technician and student of Norbert Wiener's cybernetics, woke up one day in 1950 with an epiphany: The world had so many problems, of such accelerating complexity, that humankind's only hope of mastering them was to find ways to get smarter faster. He vowed to devote his life to developing a "Framework for the Augmentation of Human Intellect." Beginning in the early 1960s under the aegis of the Stanford Research Institute, he gathered a band of researchers and began breaking conceptual and technical ground. The work culminated in a legendary public demonstration in 1968 at the San Francisco Convention Center. A video of that event is still available online, which means that today anyone can, by following some Web links and clicking a mouse, watch Engelbart introduce the world to the very idea of a link, and the mouse, and many other elements of personal computing we now take for granted.

If you watch those videos, you'll learn that Engelbart's "oNLine System," or NLS, was, among other things, a PIM. Its goal was to allow users to "store ideas, study them, relate them structurally, and cross-reference them." It provided what a computer user today would call an outliner—a program with expandable and collapsible nodes of hierarchically structured lines of information. But this outliner could be shared across a network—not only within a single office but remotely, between the downtown San Francisco auditorium and the SRI office in Menlo Park, thirty miles away, as Engelbart showed his suitably impressed 1968 crowd. Today the NLS's flickery monochrome screens and blurry typography look antediluvian, but its capabilities and design remain a benchmark for collaboration that modern systems have a tough time matching.

Engelbart showed the 1968 audience how easy it was to use NLS to make and store and share a grocery list. But the real purpose of NLS was to help Engelbart's programmers program better. In the 1962 essay that laid out his plan of research into the augmentation of human intelligence, Engelbart explained why computer programmers were the most promising initial target group. He listed nine different reasons, including the programmers' familiarity with computers and the intellectual challenge of the problems they confronted. But he also noted that "successful achievements can be utilized within the augmentation-research program itself, to improve the effectiveness of the computer programming activity involved in studying and developing augmentation systems. The capability of designing, implementing, and modifying computer programs will be very important to the rate of research progress." In other words, if NLS could help his programmers program better, they'd be able to improve NLS faster. You'd have a positive feedback loop. You'd be, in the term Engelbart favored, bootstrapping.

To Engelbart, bootstrapping meant "an improving of the improvement process." Today the term may dimly remind us that each time we turn on our computers we're "booting up." The builders of early computer systems had borrowed the term from the concept of pulling one's self up by the bootstraps to describe the paradox of getting a computer up and running. When you first turn a computer on, its memory is blank. That sets up a sort of chicken-and-egg paradox: The computer's hardware needs operating system software of some kind in order to load any program—including the operating system itself. Computer system inventors escaped this dilemma by using a small program called a "bootstrap loader" that gave the machine just enough capabilities to load the big operating system into memory and begin normal functioning. In early generations of computing, human operators would enter the bootstrap loader manually via console switches or from punch cards; modern computers store it on fixed memory chips.

For Engelbart, bootstrapping was less an engineering problem than an abstract and sometimes abstruse way of talking about the goals of his "augmentation" program. Boostrapping involved "the feeding back of positive research results to improve the means by which the researchers themselves can pursue their work." A "third-order phenomenon," it wasn't about improving a process—like, say, getting people to solve problems faster. It was about improving the *rate* of improving a process—like figuring out how you could speed up ways of teaching people to solve problems faster.

This is not a simple distinction to fathom, and that may be one reason Engelbart's project, unlike his mouse, never caught fire. Another reason, perhaps, was his determination to stick to a pure version of his "augmentation" plan. Unlike later computer innovators who elevated the term "usability" to a mantra, Engelbart didn't place a lot of faith in making tools simple to learn. The computer was to be a sort of prosthesis for human reason, and Engelbart wanted it to be powerful and versatile; he didn't want to cripple it just to ease the user's first few days or weeks in the harness. The typical office worker might be comfortable with the familiar typewriter keyboard, but Engelbart believed that the "chord keyset" he had built, which looked like five piano keys and allowed a skilled user to input text with one hand, gave users so much more power that it was worth the effort required to adapt to it. His vision was of "coevolution" between man and machine: The machine would change its human user, improving his ability to work, even as the human user was constantly improving the machine. And, indeed, as the band of researchers clustered around his Stanford lab wove the NLS into their lives, something like that could be observed. According to Engelbart biographer Thierry Bardini, "Some astonished visitors reported that [Engelbart's team had developed] strange codes or habits, such as being able to communicate in a 'weird' sign language. Some staff members occasionally communicated across the distance of the room by showing the fingers position of a specific chord entry on the keyset."

You can glean a little of that sense of weirdness today in the picture of Engelbart we encounter in the 1968 video: With a headset over his ear, one hand moving the mouse, and the other tickling the chord keyset, he looks like an earth-bound astronaut leading a tour of inner space, confident that he is showing us a better future. From the apogee of the 1968 demo, though, his project fell into disarray. He wanted to keep improving the existing NLS, whereas many of his young engineers wanted to throw it away and start afresh with the newer, more powerful hardware that each new year offered. Over time, his organization lost its sense of mission and, in the mid-seventies, foundered on the shoals of the human potential movement and Werner Erhard's est.

Engelbart's demand that users adapt to the machine found few followers in subsequent decades. As computing pioneer Alan Kay later put it, "Engelbart, for better or worse, was trying to make a violin"—but "most people don't want to learn the violin." This tension between ease and power, convenience and subtlety, marks every stage of the subsequent history of software. Most of us are likely to start with an understandable bias toward the principle of usability: Computers are supposed to make certain kinds of work easier for us; why *shouldn't* they do the heavy lifting? But it would be unfair to dismiss Engelbart's program as "user-hostile" when its whole purpose was to figure out how technology could help make exponential improvements in how people think.

Computer scientist Jaron Lanier tells a story about an encounter between the young Engelbart and MIT's Marvin Minsky, a founding father of the field of artificial intelligence. After Minsky waxed prophetic about the prodigious powers of reason that his research project would endow computers with, Engelbart responded, "You're gonna do all that for the computers. What are you going to do for the people?"

■ ■ ■

Mitch Kapor always cited Engelbart as one of his inspirations, and Agenda was in a sense a descendant of NLS. Although more modest in its ambitions and partially crippled by its dependence on DOS, it dangled a similar promise of dynamic new power to the human user who was willing to make the effort to acquire some unfamiliar skills. As Kapor and his first employees began trying to sketch the outlines of their new project, that inheritance served as one guide through the sea of possibilities that lay before them.

The decades following Engelbart's research saw a continuous revolution in computing. Machines moved out of the back office onto every desk and into every home. The graphical user interface (GUI) was developed at Xerox's Palo Alto Research Center by, among others, some of Engelbart's former colleagues and then popularized by Apple's Macintosh. The Internet connected the world's computers to form a vast, unruly global commons.

No one can deny that these novelties have changed the world. On the other hand, it's hard to argue that they have yet succeeded in providing anything like the sort of jet-boost to human capacity that Engelbart envisioned. Each computer-based advance in human productivity or convenience or creativity seems to call forth a shadow; our dreams of progress are troubled by crashes and viruses and spam. The picture of digital progress that so many ardent boosters paint ignores the painful record of actual programmers' epic struggles to bend brittle code into functional shape. That record is of one disaster after another, marking the field's historical time line like craters.

Anyone contemplating the start of a big software development project today has to contend with this unfathomably discouraging burden of experience. It mocks any newcomer with ambitious plans, as if to say, *What makes you think you're any different?*

■ ■ ■

The lead story in the *New York Times* on January 14, 2005, reported some trouble from Washington, D.C.: "The Federal Bureau of Investigation is on the verge of scrapping a $170 million computer overhaul that is considered critical to the campaign against terrorism but has been riddled with technical and planning problems." The bureau had spent a decade—and $400 million—on an ambitious program named Trilogy to modernize its computer system, replacing thirty thousand desktop computers and building a secure, modern digital communications network. But when it came to Trilogy's $170 million third part—software called Virtual Case File that would tie the new systems together, giving agents across the country ready access to the most current data and moving J. Edgar Hoover's heirs out of the paper era— the FBI hit an iceberg. An original modest plan to add some simple Web-like access to the bureau's ancient mainframe-based document storage system steadily ballooned into a master plan for a "new, collaborative environment for gathering, sharing, and analyzing evidence and intelligence data" (as it was described in the trade journal *Infoworld*). The contractor hired before 9/11 found that its client had radically expanded its wish list in the wake of that crisis. The FBI expected a finished product by December 2003, but what it got was a bug-ridden prototype. The contractor complained: Hadn't the FBI understood that when it "changed the requirements" by expanding the project, the original delivery date became impossible to meet? Maybe the FBI was having a hard time keeping up with the schedule changes because it was cycling through one chief information officer after another.

"A train wreck in slow motion," Senator Patrick Leahy called the Virtual Case File project when its debris lay spread out on a congressional committee room floor. Trilogy's woes led agency insiders to dub it "Tragedy" instead. But the plot shouldn't have been too surprising to any of the parties involved. The FBI's troubles carry echoes of a thousand and one public-sector software disaster stories. The Internal Revenue Service can match the FBI's bruises and raise the ante: It has tried three times in the last four decades to modernize its computer systems; it has had no success to date and still relies largely on a rickety mainframe system dating back to the 1960s that is held together by the digital equivalent of chewing gum and baling wire. The first attempt to upgrade IRS systems was abandoned by

President Jimmy Carter in the 1970s; the second was canceled by Congress in 1995, after ten years' labor and $2 billion failed to produce a working upgrade. The most recent effort has suffered massive delays and cost overruns. Fingers point to the usual problems: constant changes and additions to the list of requirements; limited or passive oversight by the IRS; revolving-door CIOs; unrealistic budgets and schedules.

There are countless similar stories in the annals of government software projects, from the three-decade saga of the missile defense system to the present-day Pentagon plan for a "Future Combat System" that will rebuild the U.S. Army on an "infocentric" model—a plan dependent on complex new hardware and software systems that remain in the blueprint stage ("vaporware"). Outside the United States, the picture does not look much better. In November 2004, for instance, the entire pension system for the United Kingdom was brought to its knees in what the *Guardian* called "the biggest computer crash in government history": Apparently, a simple attempt to upgrade seven computers from Windows 2000 to Windows XP was instead broadcast to tens of thousands of unprepared working machines.

But don't jump to the conclusion that government is the problem here; the record in private industry offers little solace. The corporate landscape is littered with disaster stories, too. There's McDonald's "Innovate"—$170 million down the drain in a failed effort to turn the entire fast-food chain into a "real-time network" that would give execs instantaneous reporting on the precise status of each individual batch of French fries. And there's Ford's "Everest" procurement system, a five-year, several-hundred-million-dollar black hole. It's tempting to view the multitude of monster projects gone bad as anomalies, excrescences of corporate and government bureaucracies run amok. But you will find similar tales of woe emerging from software projects big and small, public and private, old and new. Though details differ, the pattern is depressingly repetitive: Moving targets. Fluctuating goals. Unrealistic schedules. Missed deadlines. Ballooning costs. Despair. Chaos.

■ ■ ■

In 1995, a Massachusetts-based consulting firm called the Standish Group released a study of software project failures that it called the CHAOS Report. The firm has always capitalized the name as if it were an acronym, but it doesn't seem to stand for anything other than the word's original meaning. Standish surveyed 365 information technology managers at large, medium, and small companies and found that only 16 percent of their projects were successful ("completed on time and on budget, with all features and functions as initially specified"). Of the remainder, 31 percent were "impaired" or canceled—total failures. The rest, 53 percent, were considered "project challenged," a euphemistic way of saying that they were over budget, late, and/or failed to deliver all the promised features and functions.

You could quibble with the study's design and definitions, but there is no way to portray it as anything but bad news for the software industry. Standish followed up the 1995 report with regular updates, and they have shown a steady but slow improvement in the field's success rate—the most recent public number, from 2004, shows the success (29 percent) and failure (18 percent) numbers roughly flipped from 1995, but the "challenged" number (53 percent) holding steady. Whatever progress the industry has made, more than two-thirds of the time it is still failing to deliver.

Returning to the field's favorite analogy, the one between software making and bridge building, the 1995 report argued that the problem with software wasn't just a matter of too many midstream course corrections and late design changes that would never be tolerated by the builders of bridges; it was also a problem of failing to learn from mistakes: "When a bridge falls down, it is investigated and a report is written on the cause of the failure. This is not so in the computer industry where failures are covered up, ignored, and/or rationalized. As a result, we keep making the same mistakes over and over again."

In fact, though, there is an entire software disaster genre; there are shelves of books with titles like *Software Runaways* and *Death March* that chronicle the failures of one star-crossed project after another. These stories of technical ambition dashed on the shoals of human folly form a literature that is ultimately less interesting as fact than as mythic narrative. Over and over the saga is the same: Crowds of ardent, ambitious technologists march off to tackle some novel, thorny problem. They are betrayed by bad managers, ever-changing demands, and self-defeating behavior. They abandon the field to the inevitable clean-up squad of accountants and lawyers.

The genre's definitive work to date is *The Limits of Software*, a disjointed but impassioned book by an engineer named Robert Britcher. Britcher is a veteran of big software's ur-disaster, the train wreck against which all other crack-ups can be measured: the Federal Aviation Administration's Advanced Automation System (AAS). AAS was a plan to modernize the air traffic control system; it began in 1981 and "terminated" in 1994 after billions of dollars had been spent, with virtually nothing to show.

Britcher, who had worked on the original air traffic control system built by IBM in the 1960s and then labored in the trenches on its planned replacement, writes that AAS "may have been the greatest debacle in the history of organized work." The FAA laid down a set of rules for its new system: Every bit of software created for it must be reusable. The entire system had to be "distributed," meaning that each air traffic controller's workstation would synchronize its information in real time with all the others. The cut-over to the new system had to be seamless, and all changes had to be made while the existing system was running. (This was "like replacing the engine on a car while it is cruising down the turnpike.") Paper printouts might help that process—air traffic controllers relied on them when the old computer system got flaky—but the FAA forbade printers: They were too old-school. And on and on.

At its peak, the AAS program was costing the government $1 million a day; two thousand IBM employees labored on the project, producing one hundred pages of documentation for each line of program code. But in the end, Britcher concludes, "The software could not be written." The FAA's demands were simply beyond the capacity of human and machine. "If you want to get a feel for how it was," he writes, "you can read the *Iliad*." With its sense of tragic inevitability, of powerful men trapped by even more powerful forces, *The Limits of Software* is itself a kind of digital *Iliad* populated by doomed desktop warriors. In the programming campaigns Britcher chronicles, the participants—not so much overworked as overtaxed by frustration—smash their cars, go mad, kill themselves. A project manager becomes addicted to eating paper, stuffing his maw with bigger and bigger portions at meetings as the delays mount. No one—including, plainly, the author—escapes scar free.

> One engineer I know described the AAS this way. You're living in a modest house and you see the refrigerator going. The ice sometimes melts, and the door isn't flush, and the repairman comes out, it seems, once a month. And now you notice it's bulky and doesn't save energy, and you've seen those new ones at Sears. So it's time. The first thing you do is look into some land a couple of states over and think about a new house. Then you get I. M. Pei and some of the great architects and hold a design run-off. This takes a while, so you have to put up with the fridge, which is now making a buzzing noise that keeps you awake at night. You look at several plans and even build a prototype or two. Time goes on and you finally choose a design. There is a big bash before building starts. Then you build. And build. The celebrating continues; each brick thrills. Then you change your mind. You really wanted a Japanese house with redwood floors and a formal garden. So you start to re-engineer what you have. Move a few

bricks and some sod. Finally, you have something that looks pretty good. Then, one night, you go to bed and notice the buzzing in the refrigerator is gone. Something's wrong. The silence keeps you awake. You've spent too much money! You don't really want to move! And now you find out the kids don't like the new house. In fact, your daughter says "I hate it." So you cut your losses. Fifteen years and a few billion dollars later, the old refrigerator is still running. Somehow.

■   ■   ▮

If programmers paid too close attention to the legacy of software disasters past, nothing would ever get coded. The likelihood of failure would be too daunting. We would just have to cross our fingers and pray that our old refrigerators wouldn't conk out. But except for haunted burnouts like those FAA veterans that Britcher memorializes, most participants in the creation of new software are either blissfully ignorant of the past or recklessly confident of the future—blithely certain that this time things will be different.

"All programmers are optimists," Frederick Brooks wrote in 1975. "Perhaps the hundreds of nitty frustrations drive away all but those who habitually focus on the end goal. Perhaps it is merely that computers are young, programmers are younger, and the young are always optimists." But programmers' innate optimism is often obscured by their forthrightness about difficulties and problems. Reporters know that when you can't get a straight answer from a salesperson or a marketer about a product, an engineer will (when not locked away or ordered to be silent) speak the hard truth. In my very first conversation with Chandler's original software architect, John Anderson, at a time when I understood the project to be in its infancy, he sat down across from me and, before I could ask a question, declared, "Here's what our three biggest failures are." So if programmers are optimists by nature, they also have a keen eye for the downside. A hyperactive imagination for disaster scenarios is a professional asset; they have to think through everything that can go wrong in order to practice their craft.

If you want to change the world, the Italian radical Antonio Gramsci famously declared, you need "pessimism of the intellect, optimism of the will." Today's software creators are improbable heirs to that binary mind-set.

As he started imagining the project that would become Chandler, Mitch Kapor was certainly an optimist. It was a dark winter for the software industry, the bottom of a down cycle. But he had the financial wherewithal to follow his bliss, the eagerness of the successful entrepreneur to roll up his sleeves, and, of course, an idea of what he and the world needed.

When I first approached Kapor to discuss writing about his work, he told me that Chandler would make a bad case study. "We're atypical," he said. "We're open source, we're nonprofit, and we're taking the design-first approach of the 'Software Design Manifesto' [an essay Kapor had written in 1990]. We're trying to practice what we preach."

He was, in a sense, right. Chandler lacked many of the usual constraints of software projects. There was no corporate hierarchy standing behind its deadlines; there were no impatient investors sitting offstage, ready to pull the plug if the code didn't ship. But that also made the case look, to me, more valuable. In observing Chandler's progress, we could factor out most of the money side of things; the project's fate would not be shaped by the financial pressures that so often take the blame for software disasters. Chandler offered a look at the technical, cultural, and psychological dimension of making software, liberated from the exigencies of the business world. If there are barriers inherent in the very nature of programming that prevent us from getting what we want out of software, perhaps we would glimpse them here. And if Kapor and his team emerged with new insights into how to overcome those barriers, even better.

But in saying Chandler was atypical, Kapor was also invoking the mantra that project leaders everywhere offer to try to clear the air of paralyzing memories: *This time will be different.* He was flinging his optimism in the face of history—not just the industry's, but his own. On Technology, the company he had started after leaving Lotus and finishing Agenda, had turned out, well, not so different. Kapor had started in 1987 with high ambitions to make PC software smarter—building Agenda's ability to recognize dates and some names into a broader, open product, one that would work on all different kinds of computers, enabling groups of office workers to collaborate. Though no one ever seemed quite certain exactly what On was building, clearly it would be impressive. But after a year and a half with virtually nothing to show, "We took the plan out and we shot it," Kapor told the *Wall Street Journal.*

Cutting the size of the company in half, he slimmed down On's goals and handed the new plan to a small team. Two years later, On shipped its first product: an add-on to the Macintosh operating system that made it easier for users to organize and find files. The company went on to release two more software products: a tool for setting up meetings over computer networks, and a program for collaboratively editing a shared memo. The products were critical hits but failed to catch fire, and they were limited to the Macintosh, which had become a narrow, isolated market. Meanwhile, On Technology had become one more chapter for the software disaster tomes.

Kapor was right: Chandler *was* atypical. But what project wasn't?

"There's no such thing as a typical software project," Andy Hertzfeld liked to say. "Every project is different."

CHAPTER 3

PROTOTYPES AND PYTHON

[2001–NOVEMBER 2002]

H ow do you organize a music collection? If you have a mountain of CDs, unless you're content to leave them in a state of total disorder, you have to pick one approach to begin with: alphabetical by artist, maybe. Or you'll start organizing by genre—rock in one pile, jazz in another, classical in another—and then sort alphabetically by artist within those piles. If you're more unorthodox, maybe you'll order them alphabetically by title or by label or by year of release. Or maybe you just let them accumulate according to when you bought them, like geological strata layered in chronological sequence.

Once you pick an approach to filing your albums, though, you're committed; switching to a different scheme requires more work than most of us are willing to invest. (In 2004, a San Francisco conceptual artist decided to re-sort all the books in a used bookstore according to the color of their bindings. He wrote that the effort required "a crew of twenty people pulling an all-nighter fueled by caffeine and pizza.") That's a limitation of the world of physical objects. But there are advantages, too. Once you've lived with your arrangement a while, you discover you can locate things based on sense memory: Your fingers simply recall that the Elvis Costello section is over at the right of the top shelf because they've reached there so many times before.

As music collections have begun to migrate onto our computers, we find that we've gained a whole new set of possibilities: We can instantly reorder thousands of songs at a momentary whim, applying any number of criteria in any combination. Our music software even lets us view our collections through previously unavailable lenses, like "How many times have I listened to this before?" We've gained the same power over our troves of recordings that we get any time we take something from the physical world and model it as data. But we've also lost some of the simple cues we take for granted in the physical world—the color of the CD box's spine that subliminally guides our fingers, or the option to just toss a box on top of the CD player to remind yourself that that's what you want to hear next.

When we move some aspect of our lives into software code, it's easy to be seduced by novel possibilities while we overlook how much we may be giving up. A well-designed program will make the most of those new capabilities without attempting to go against the grain of the physical world orientation that evolution has bequeathed us. (I remember the location of this button from yesterday because my brain is wired to remember locations in space, so it had better be in the same place tomorrow!)

Too often, though, we end up with software that's not only confusingly detached from the world we can touch and feel but also unable to deliver on the promise of flexibility and versatility that was the whole point of "going digital" in the first place. You want to organize your music by genre? Go ahead, but please don't expect to add to the list of genres we've preloaded into the program. You're an opera fan and need a different scheme for sorting files than the one that works for the Eagles and Eminem? Sorry, we baked the Artist and Song Title categories so deeply into the software's structure that you'll need a whole new program if you expect to track composers and singers and arias.

This happens not because programmers want to fence you in, but because they are so often themselves fenced in by the tools and materials they work with. Software is abstract and therefore seems as if it should be infinitely malleable. And yet, for all its ethereal flexibility, it can be stubbornly, maddeningly intractable, and it is constantly surprising us with its rigidity.

That paradox kicks in at the earliest stages of a programming project, when a team is picking the angle of attack and choosing what languages and technologies to use. These decisions about the foundations of a piece of software, which might appear at first to be lightweight and reversible, turn out to have all the gravity and consequence of poured concrete.

■ ■ ■

In the spring of 2001, Mitch Kapor, Morgen Sagen, and Al Cho began meeting regularly to talk about the software they were going to build. During his last year at Excite@Home, as the company began to come apart at the seams, Sagen had built for his own needs and amusement a little Web-based program to make bookmarks and other random personal information easy to find and sort. He drew on that experience, just as Kapor was drawing on Agenda, as they began asking deceptively simple questions: How do we organize information? How would we model that organization? What structure would we need to build to share the answer to that question with a computer?

The meetings stretched on; the whiteboards filled up.

Because it so often takes ages before a new piece of software is ready for anyone to use, programmers often test their assumptions against "use cases"—hypothetical scenarios about how imaginary people might need or want to use a program. One use case for the OSAF project that emerged early on, and would keep reappearing, was organizing a big personal music collection. The example came naturally; in his youthful *Wanderjahre*, Kapor had done time as a DJ. And at that time in 2001, Apple's now-popular iTunes program had only just been born and was limited to Macs; the iPod, its companion device, did not yet exist.

Sagen and Cho began to sketch out what kind of data structures a program would require in order to meet that and other use cases. What kinds of categories and labels would you need? How would those categories relate to one another? How would the basic data (say, the actual files containing music) relate to all the metadata, the artists' names and album titles and song numbers that describe the music? How would you efficiently store everything? And was there a way of doing it all that didn't lock you in, that let you keep adding new kinds of metadata as you went along?

After a certain number of weeks and full whiteboards, they came up with a broad answer. You could model just about anything in a simple three-part format that looked something like the subject-verb-object arrangement of a simple English sentence:

```
<this> <has-relationship-with> <that>
```

Then they discovered that the answer they'd come up with had already been outlined and at least partially implemented by researchers led by Tim Berners-Lee, the scientist who had invented the World Wide Web a dozen years before. Berners-Lee had a dream he called the Semantic Web, an upgraded version of the existing Web that relied on smarter and more complex representations of data. The Semantic Web would be built on a technical foundation called RDF, for Resource Description Framework. RDF stores all information in "triples"—statements in three parts that declare relationships between things. This was very close to the structure Sagen had independently sketched out, with the advantage that a considerable amount of work over several years had already been put into codifying the details.

"Good programmers know what to write," Eric Raymond had written in "The Cathedral and the Bazaar." "Great ones know what to rewrite (and reuse)." There was no need to reinvent the RDF wheel; maybe OSAF could just hitch a ride on it.

▌ ▐ ▌

The music library use case didn't stay hypothetical for very long; the next programmer to join OSAF embodied it. Andy Hertzfeld had made a name for himself thanks to his central contributions to the original operating system for Apple's Macintosh in 1984, which established the course of personal computing for the next two decades and set a standard for innovation and elegance that every new project in Silicon Valley still aspires to. In the years since he had left Apple, soon after the Macintosh's introduction, Hertzfeld contributed his technical skill, boyish enthusiasm, and imagination to a couple of ambitious failures—first, General Magic, a start-up company that aimed to revolutionize the market for handheld "digital assistants," and then Eazel, an open source company that ran out of money in the dot-com collapse before it could complete the programs it had designed for Linux users.

Hertzfeld had become a fervent and vocal convert to the open source gospel; on the same day that Eazel shut down, Kapor, who was trying to decide whether to go open source with his new project, came to visit him and told him about his plans for OSAF. A few months later Hertzfeld decided to join Kapor's project as a full-time volunteer. In addition to being a software legend, he was also a serious music collector with a stockpile of several thousand digitized recordings and a particularly avid dedication to the metamorphic catalog of Bob Dylan. In other words, he was a walking use case.

Hertzfeld began joining OSAF's weekly chalk talks and soon began to build Vista, a working prototype of the kind of flexible program Kapor had in mind. Hertzfeld would go off, write some code, and then return to Kapor's office to show off the new tricks Vista could do.

Vista was a "user-facing" program: It took care of displaying data to a user and shaping how you could input, organize, and change that data. It stored all its information in Shimmer—a database, or, as it came to be called in the OSAF world, repository, that Morgen Sagen had built on an RDF foundation. Together, Vista and Shimmer offered a glimpse of the kind of breakthrough Kapor dreamed of: a tool that let you manage your life's details from within one program according to rules set by you rather than the computer. But it was limited, and it crashed a lot. It was a true prototype, designed to be milked for whatever lessons it could teach and then discarded.

In any project that is introducing a new technology or design, Frederick Brooks had advised, "plan to throw one away," because you almost certainly won't get it right the first time. All you can control is whether you actually *plan* to get it wrong, or rashly "promise to deliver a throwaway to customers."

■ ▪ ▫

By the spring of 2002, Kapor had a clearer picture of the program he wanted to build. Like Agenda, it would be a PIM, or personal information manager, a tool for managing email, appointments, addresses, tasks, and notes. But it would be "cross-platform": You could run it whether you were using Windows, Mac, or Linux, the three most popular personal computer operating systems of the day. And it would be explicitly designed so that any developer could add new capabilities to it. Open source programmers could code new modules or "parcels" for, say, managing digital photos or music collections. And nontechnical users would be able to add new categories and labels to the program on the fly.

These were ambitious goals. Kapor began hiring developers. Katie Parlante, a studious Stanford-trained software developer who had spent the dot-com years at two boom-and-bust start-ups, was the first addition. Parlante had a reflective presence and an even keel, and she sometimes seemed to hang back from the fray of OSAF's disputes, but she had a sharp technical mind, and when she did speak up, people listened. Behind her quietness, she was persevering, and over the years she would steadily move toward the center of OSAF's management. But for now she was happily researching user behavior patterns, exploring ways a PIM could help people avoid the tedious game of telephone tag, and compiling the most common complaints users have about their calendar software.

Shortly after Parlante joined OSAF, Kapor hired a veteran programmer named John Anderson as OSAF's systems architect. Defining that title is not simple. I once sat in a banquet room at a conference for military software contractors and watched as Raytheon's vice president of engineering asked the crowd how many considered themselves "systems architects." Half the audience raised their hands. Another third identified themselves as "systems engineers." The Raytheon exec then asked them to define "systems architect" and distinguish it from "systems engineer." A long silence followed and then a series of faltering, unsatisfying stabs at an answer. Some in the crowd seemed affronted, as if the very question were impertinent. Finally, one attendee stood up and started reciting the verbiage from an IEEE (Institute of Electrical and Electronics Engineers) standards document, whose boilerplate bureaucratic language only underscored the speaker's point: that a room filled with professionals whose work depended on precise specification could neither compose nor agree upon a simple, unambiguous definition of their own job titles.

At OSAF, John Anderson's systems architect title meant that he held chief responsibility for choosing the building blocks that the programmers would use to create their software: Which programming language should they code in? What tools should they use to create the program's graphical interface—the windows and menus and dialogue boxes that users would actually see? What kinds of software technologies should they use to store the program's data? What standards for data exchange should they adopt? There were a thousand choices to make, and a bewildering array of options for each choice.

Anderson has a Ph.D. in zoology and a lifetime's experience building software, especially for the Macintosh and Next platforms. His management stint at Next had left him with some bruises. "Frankly, I probably did kind of a bad job as a manager. I didn't really know what I was doing," he now says. "But I learned a lot. In the past, I always thought the best idea won because it was the best idea. And what I realized was, there's actually politics. Believe it or not, that was a new experience." Most recently he had become a highly regarded independent consultant, parachuting into companies with particularly thorny technical problems. "Being a contractor is kind of like being a houseguest in a dysfunctional family. You get a chance to leave when it's over," he says. "And you never build up all those emotional tangles and tugs-of-war you feel when you're really part of an organization. I got to explore a lot of interesting problems on my own."

Anderson was used to having a steady flow of contract work, but he hit a dry patch at the start of 2002, when the dot-com downturn cascaded through the tech-industry economy. He started calling friends, looking for leads. Andy Hertzfeld told him about OSAF, and, he recalls, "It was the first project that came along that seemed as fun as the project I was working on on my own. And it was at the funnest stage—everything was so early that nothing was screwed up."

For programmers, just as for writers and artists and everyone whose work involves starting with a blank slate, the "funnest" time of a project often falls at the very beginning, when worlds of giddy possibility lie open, and before painful compromises have shut any doors.

"The programmer, like the poet, works only slightly removed from pure thought-stuff," Frederick Brooks wrote. "He builds his castles in the air, from air, creating by exertion of the imagination. Few media of creation are so flexible, so easy to polish and rework, so readily capable of realizing grand conceptual structures."

Brooks's passage carries echoes of some of Shakespeare's best-known lines from *A Midsummer Night's Dream*:

> *The lunatic, the lover, and the poet*
> *Are of imagination all compact . . .*
> *The poet's eye, in a fine frenzy rolling,*
> *Doth glance from heaven to earth, from earth to heaven;*
> *And as imagination bodies forth*
> *The forms of things unknown, the poet's pen*
> *Turns them to shapes, and gives to airy nothing*
> *A local habitation and a name.*

The words belong to Theseus, the Duke of Athens, who lumps together "the lunatic, the lover, and the poet" as victims of their own imagination's flights. A devotee of "cool reason," he's contemptuous of the labor he describes—the "tricks" of "seething brains."

But we may hear his words differently: They offer a precise outline of what any creative work requires in the process of making the imagined real. Programs are indeed "only slightly removed from pure thought-stuff," but they cannot remain thought-stuff or they will not do anything. Programmers must take an idea from the realm of thought and embody it in functional lines of code—assigning it "local habitations and names" in the world of the computer.

■ ■ ■

The memory banks of the earliest commercial computers, in the 1950s and '60s, were built out of wound wire coils known as ferrite cores. Ever since, even as the industry abandoned electromagnets for semiconductor chips, programmers have referred to the computer's active memory as "core." In many systems, when something goes wrong, when the machine freezes up because of an irreconcilable conflict or unexpected error, there is a "core dump"—the computer drops everything, grinds to a halt, and spits out a file reporting the exact contents of its memory at the moment of failure, offering bug hunters a heap of clues to dig through.

The core is a physical entity, not just a metaphor. It has local habitations and names. Your computer's memory is a vast array of actual locations—however microscopic they may be in today's advanced chips—and those locations have names and addresses. At any given instant each address in that memory, each bit, will contain a one or a zero; it is a simple switch, and it will be on or off. Everything else that happens in a computer, whether it is keeping a corporation's books or running a *Doom* death-match, is built on that binary foundation—as the central processor chip, or CPU, following a programmer's instructions, takes those bits from memory, manipulates them, and then puts them back.

Software is different; it has no core. It is onionlike, a thing of layers, each built painstakingly and precariously on the previous one, each counting on the one below not to move or change too much. Software builders like to talk about laying bricks; skeptics see a house of cards. Either way, there's a steady accumulation going on. New layers pile on old.

Programmers call these accretions "layers of abstraction," because each time a new one is added, something complex and specific is being translated into something simpler and more general. The word *abstraction* comes from the Latin for *draw away*; here's one computer science definition of the term: "The process of combining multiple smaller operations into a single unit that can be referred to by name." Abstraction begins in the nursery. It is what children learn when they realize that if they want an apple, they don't have to find one to point to but can simply say the word that refers to it. They also learn that abstraction depends on a common language, on shared assumptions of what class of objects the word *apple* refers to.

"This is what programmers do," wrote Eric Sink, a programmer who led the creation of the Web browser that became Microsoft's Internet Explorer. "We build piles of abstractions. We design our own abstractions and then pile them up on top of layers we got from somebody else." And every year the piles grow higher.

■ ■ ■

At the very bottom of the pile, sitting right on top of the core memory, is assembly language, invented half a century ago to make it easier for programmers to manipulate the core's zeros and ones by hand. Instead of simply writing lots of binary code, like 10110000 01100001, they could write the equivalent assembly language instruction, which uses commands with names like "mov" and gives variables easier-to-remember labels. You're still pushing bits around by hand, but you have some easier handles to grab.

For most human beings assembly language is difficult to learn and arduous to write. In the memorable phrase of programmer-essayist Ellen Ullman, it's "close to the machine"; if it's not right inside the machine's head, it's whispering in its ear. It's also not very adaptable; there's a different assembly language for each type of CPU, requiring you to rewrite your code each time you want to run it on a different make of computer. So the programmers of yore started adding more layers of abstraction, with each layer moving a little further away from the machine and closer to the human being.

The decade from the mid-1950s to the mid-1960s saw a sort of Cambrian explosion of new programming languages, including many that are still in use or that became the ancestors of today's workhorses: Lisp, Cobol, Algol, Basic. (These language names are technically all acronyms, and you'll often find them in all upper case, but I'll spare your eyes.) The very first to achieve wide use was Fortran—short for "FORmula TRANslating system." In Fortran, laborious sequences of assembly language procedures were summarized in brief commands. The human programmer would write a sequence of these commands—source code; then a kind of uberprogram running on the computer called a compiler would translate those commands into object code in the machine's own language.

It was as if people who had communicated by spelling out every word to one another, letter by letter, suddenly figured out how to talk in syllables.

The leap was vast—so vast that at first Fortran's promoters believed they had not so much advanced computer programming as entirely transcended it. Their invention, they wrote, would "virtually eliminate coding and debugging." They touted Fortran as an "automatic coding system," since it replaced the opaque binary code of machine language or the only slightly more accessible assembly language with a relatively more readable sequence of commands like READ, ASSIGN, GOTO, and STOP. Fortran seemed at first to offer a view of a promised land where the hard struggle to bend computers to human wishes—"hand to hand combat with the machine," as Fortran inventor John Backus described the work of primitive coding—vanished, replaced by the magic of a system that let people communicate their wants and needs directly to the machine, in their own language.

But like so many innovations in programming over the following decades, it could not deliver on its dream. Ultimately, Fortran, for all its originality and utility, solved one set of problems only to introduce others. Within a decade of its introduction, the entire field of programming was the victim of its own success, and experts gathered in international conclaves to ponder a way out of "the Software Crisis."

█ █ █

One reason Fortran prospered was that it defied the experts' prediction that the code of such a higher-level language would inevitably run more slowly than the handcrafted machine code composed by human beings. As you moved up the pile of layers, further from the machine, you asked the computer to do more work; with Fortran you had to run your source code through the compiler before the computer could run your program. But once you did, the resulting code turned out to be remarkably efficient: The programs that the Fortran compiler produced "automatically" proved just as good as the programs people coded directly in machine or assembly language. For computer scientists, that revealed the Fortran compiler to be a thing of marvelous efficiency, verging on beauty. It turned out to cost only a little extra computer time to run Fortran programs, and it saved a lot more programmer time.

In the era of Fortran's invention, computer time was costly; the machines were few and big and fantastically expensive. But that was already beginning to change. At about the same time that Fortran was taking off, in 1965, Intel Corporation founder Gordon Moore propounded his celebrated law plotting the rapid growth in the power of computer hardware: The number of transistors you could pack on a computer chip, Moore noticed, increased exponentially, doubling every year or two. Although Moore tweaked the exact figure a couple of times, the essential point never varied. Any number that doubles regularly over time gets really big surprisingly soon. Doubling the number of transistors per chip meant you could process computer instructions twice as fast or at half the cost. Thanks to this phenomenon, throughout the four decades following Moore's observation, you could count on computer hardware to get faster and cheaper every year.

For programmers this meant that the habits of computing's youth—when you had to fine-tune or "optimize" your code to use as little memory and as few "cycles" (units of processor time) as possible—were not necessarily the best instincts to carry into the future. Over the years, certainly by the time Kapor and OSAF began their work, it became clear that processor time was cheaper than programmer time. Many serious software developers—like writers who make every word count or cooks who find ways to use every last scrap of a carcass—still maintain the virtue of frugality and discipline. But when it comes down to a trade-off between making the computer work harder or making the programmer work harder, as it so often does, the choice couldn't be easier to make.

That's because there is no Moore's Law for software. Chips may double in capacity every year or two; our brains don't.

■ ▮ ▯

When John Anderson, Andy Hertzfeld, and Mitch Kapor sat down in the summer of 2002 to select the programming language OSAF would use, they did not need to remind one another of this history. It was part of what they had experienced over the previous three decades in the personal computer industry. In the years since Fortran's invention, a veritable Babel of programming languages had emerged, each with strengths and weaknesses. Some were better suited for business applications, others for science; some were good for quick and dirty jobs, others for fulfilling grand ambitions. Each successful language had its cadre of evangelists. The arguments between them often resembled rifts between religious sects, and even the issues at stake had the flavor of doctrinal disputes. The central conflicts in programming language design have always been Miltonic dilemmas, choices between free will and predestination. For instance: Should the language give programmers the power to poke directly into the computer's memory—along with the freedom to make machine-crashing mistakes? Or should the language create zones of safety that limit the possibility of error—at the cost of tying the programmer's hands?

For the Vista prototype, Hertzfeld had used a language called Python, invented in the late 1980s by a Dutch programmer named Guido van Rossum who named it in honor of Monty Python's Flying Circus, the British comedy troupe. (Monty Python's form-smashing absurdism has always found some of its truest fans in computer labs; we call the flood of unsolicited and unwanted email "spam" thanks to the Internet pioneers who, looking to name the phenomenon, recalled a Python routine featuring a luncheonette menu offering nothing but variations on "eggs, sausage, spam, spam, spam, and spam.")

Python is an interpreted language. Where compiled languages run programmers' source code through a compiler ahead of time to translate it into machine-readable binary code, interpreted languages perform that translation when you run the program. The source code gets translated line by line by the interpreter and fed to the processor for execution. This makes interpreted languages less efficient, since you're always running two programs at once, the program you want to use and the interpreter. But it also makes them nimbler. The programmer doesn't have to wait for each code change to be run through the compiler to see whether it worked, and the program has the flexibility to respond to changes on the fly. It's like the difference between carrying a pretranslated guidebook on your trip to Slobovia or bringing along a human interpreter. Interpreters add a lot more to your overhead, but if there's an earthquake or an insurrection, having that Slobovian expert at your elbow is going to be a lot handier.

Interpreted languages like Python were the object of a certain disdain from many software developers, who viewed them as "scripting languages"—tools for cranking out short program scripts, useful for whipping up some quick utility code, inadequate for serious work. But by 2002, Python advocates were beginning to challenge that assumption.

Anthony Baxter, an acerbic Australian programmer who served as a manager for new releases of Python, introduced talks about a Python-based Internet telephony project with a slide that read: " 'Scripting Language' my shiny metal arse!" (He was paying homage to a favorite line of Bender, the grumpy robot in Matt Groening's *Futurama* series.) Andy Hertzfeld, too, was singing Python's praises. He liked to say that he was three times more productive using Python than using Java, which in turn had made him three times more productive than he'd been when he used C, the language most programmers in the personal computing industry or the university world had cut their teeth on. The Python interpreter itself is written in C; that makes it another layer, a higher-level language, one more step up the ladder, further from the machine and closer to the human being.

Typically, code in higher-level languages like Python—or its longtime rival, Perl—can be tighter and less verbose than code written at a lower level; more stuff has been abstracted. Van Rossum says that a Python program can usually accomplish the same task as a C or C++ program using three, five, or even ten times less code. That doesn't only mean that there's less for a programmer to write and type; it means that when programmers need to go back and add a new feature or fix a bug, it's easier for them to figure out what the code does. "I'm convinced that a single programmer has a certain amount of code that they can handle in their heads, or in a file, that they can comfortably manage," van Rossum says. "And the amount of code is expressed roughly in lines of code, not in the complexity of what the code actually does. So if there's five times as much code, then it really takes five times as much effort to maintain."

Despite his whimsical choice of a name for his language, van Rossum is mostly a no-nonsense pragmatist. The spirit of the Monty Python fan slips into his conversation only rarely, as when he introduces himself to an audience of programmers with a biographical slide that reads:

Age 4:   First Lego kit
Age 10: first electronics kit
Age 18: first computer program (on punched cards)
Age 21: First girlfriend :-)

Just as dogs often come to resemble their owners, it seems that programming languages end up reflecting the temperaments and personalities of their creators in some subtle ways. For instance, Larry Wall, the linguist who is the father of Perl, is a droll spirit enamored of verbal and visual puns. At a conclave for open source programmers in 2004, delivering an annual review of new Perl developments, he found a way to connect a meditation on the mutating visual geometries of screensavers with an account of his recent battle with a stomach tumor, then tied it all together with a vision of the community of Perl programmers "performing random acts of beauty for each other."

So it's no surprise to learn that Perl itself is a language that prizes flexibility of expression above all; one of its slogans is "There's more than one way to do it." (Acronym-happy programmers, adding one more layer of abstraction, boil this down to TMTOWTDI.) Those less in love with Perl argue that the freedom it offers to achieve the same result any number of ways also makes it harder for programmers to read one another's code. With Python, van Rossum says, it's TOOWTDI: There's only one way to do it. In one observer's characterization of the two languages' partisans, Perl-ites are "chaotic/good trickster archetypes" where "Pythonistas are peace-ful, have-their-glasses-on-a-little-string types, like hobbits or the Dutch."

The Perl and Python crowds have kept up a heated rivalry for years, but their bets about relative performance and occasional outbreaks of pie-throwing are more playful than bitter. Van Rossum and Wall met for the first time in the mid-1990s. "The Perl and Python communities were already head-butting, but Larry and I got along just fine," van Rossum says. "Neither of us was particularly concerned or worried about what would happen if the other language would 'win.' Neither of us had much ego invested in winning. We didn't do that kind of competition."

Languages like Perl and Python came into their own during the Inter-net boom of the later 1990s, when companies faced with tight deadlines turned to these "scripting languages" as the "glue code" or "duct tape" to strap Web servers and databases together into functioning Web sites. But they were still considered lightweights. Some programmers, for instance, turn up their noses at Python's use of white space: Where many languages use brackets or other symbols to mark off discrete blocks of code, Python simply uses indents. That makes Python code easy on a programmer's eyes, but it can also be fragile, since indents tend to vanish in email or change when you move code from one operating system to another.

Another aspect of Python that caused critics to sniff was the relatively esoteric question of its approach to "typing" variables. Programming variables are similar to, but not exactly the same as, those x's and y's that filled the blackboards of our algebra classrooms: They are placeholders for data, names for slots in the computer's memory that will be filled with information a user will input (or another part of the program itself will output). When you type in your starting balance in a banking program, for instance, the program assigns it to a variable with a name like *balance.starting*. The program assigns that variable a value at some point, maybe by asking you. It then stores the value in the computer's memory, and the next time the program code refers to *balance.starting*, the computer will retrieve that value.

Some programming languages are designed to be relatively strict about requiring programmers, when they create a new variable, to declare in advance what type it is: A simple integer? A floating point decimal number? An alphanumeric string containing any kind of character? Other languages are more permissive. Some languages enforce these rules at compile time, when the source code is run through the compiler; others don't enforce them until run time, when the program is executed. The terms *static* and *dynamic typing*, and *strong typing* and *weak typing*, describe these approaches, but they are ambiguously used and widely misunderstood; they are themselves, as it were, weakly typed. A computer scientist named Benjamin Pierce, who is the author of whole books on the subject, once wrote, "I spent a few weeks trying to sort out the terminology of strongly typed, statically typed, safe typed, etc., and found it amazingly difficult. The usage of these terms is so various as to render them almost useless."

Despite or perhaps because of this confusion, programmers' mailing lists and forums are overloaded with heated arguments on the subject. These distinctions may seem a small matter, but in the land of code they are the stuff of quasi-religious war. Python's detractors looked down their noses at what they view as its relative laxity on typing; the language's partisans argued that it makes the trade-offs in this area more intelligently than its competitors. Either way, its interpreted nature ensured that it would be considered a less serious language by many. But in a couple of other major respects, Python more closely resembles industrial-strength languages than the scripting languages with which it got lumped. First, it provided "garbage collection": When the program, which claims a space in memory for each variable, no longer needs the space, Python, like Java and other languages of recent vintage, takes care of freeing it up. (Many older languages required you to manage memory by hand, turning the programmer into a sort of hotel clerk for the memory's temporary informational lodgers.) Also, Python, like Java and C++, was "object oriented."

Object-oriented programming: More than any other of the arcane terms that have migrated beyond the closed world of the programmer, this one has resisted clarity and invited misunderstanding. Even after reading hundreds of pages on the subject, discussing it with experts, and attending a number of lengthy lectures on the topic, I cannot say with 100 percent confidence that my own explanation of the phrase will satisfy every reader. Still, here goes: Object-oriented techniques organize programs not around sequential lines of commands but instead around chunks of code called objects that spring to life and action when other objects call on them. Objects relate to other objects via strictly defined inputs and outputs, so that programmers writing code that must interact with them need not concern themselves with what's happening inside them. Object-oriented programming introduced a whole new vocabulary to the field. Software objects interact with one another by sending "messages" and triggering "events." Typically, objects can be organized into "classes" that share traits, and they can inherit characteristics from their "parents." (This terminology sometimes gives discussions of object-oriented programming the strange ring of Marxist theory cut with genetic science.)

The history of object-oriented programming stretches back to the 1960s, but the approach didn't sweep the market until the late 1980s, when programmers began to move from the sequential world of the command-line interface—a literally linear environment, in which the user's interaction with the computer happened line by typed line—to the anarchy of the graphical interface, in which the user might click anywhere ("initiate an event") at any moment, and the program, like an attentive goalie, had better be ready. The principles of object-oriented programming looked like an ideal tool for managing this chaotic world.

Before the fad peaked, the technology press had latched onto the notion of objects and oversimplified it to mean, simply, "bits of reusable code." Software objects—like Fortran and so many other previous innovations—would change everything, reducing programming itself from a monumental undertaking to a simple chore of borrowing components from libraries. But in practice, though the object-oriented approach gave programmers a leg up as they constructed ever more complex edifices of code, it did not open a road to the marketers' utopia where programmers could write components once and reuse them anywhere. There was more and more code around: libraries of objects and functions and modules, increasingly available under open source licenses that meant you could tinker with them and incorporate them for free. Programmers setting out on big new projects could stand on the shoulders of ever taller giants. But they still had · to decide where they wanted to go.

As Frederick Brooks put it in his "No Silver Bullet" essay: "The hardest single part of building a software system is deciding precisely what to build."

▌ ▌ ▌

Python appealed to Kapor and company for many reasons. Like the program they aimed to build, it was open source and cross-platform (the same Python program could run on Windows, Macintosh, and Linux computers). Some newer alternatives to Python had their allure—one called Ruby, invented by a Japanese programmer, was winning a lot of attention—but Python had the advantage of maturity. This meant not only that it had long ago ironed out many bugs and streamlined its performance, but also that programmers had built piles of tools for it—new layers and helper programs so that the OSAF coders wouldn't have to start from scratch. As Python devotees liked to put it, the language came with "batteries included." Python's trade-offs between developer time and machine time meant that it was certainly not as fast as some other languages, but it allowed you to find bottlenecks in the code and optimize them. You could drop down "closer to the machine," rewrite those performance-limiting sections in the C language, and get the best of both worlds. And Python enthusiasts talked with awe of "Guido's time machine": It seems that often, when programmers complain of a particular shortcoming in Python, van Rossum points out that the most recent update of the language has already solved the problem—giving rise to the joke that he's able to travel backward in time to fix his creation.

For every decision in the technology industry there is a safe choice—an embodiment of the old saying, "Nobody ever got fired for buying IBM." The "no one ever got fired" choice of programming language for OSAF would have been Java. It satisfied many of the needs that Kapor's team had identified: It was cross-platform and object-oriented; business people were comfortable with it; and programmers knew it well. But it was not truly open source (Sun Microsystems owned it), and Kapor's experience building a modern version of Agenda in Java left him doubtful that it could build the sort of beautiful, intuitive graphical user interface he dreamed of.

Python wasn't perfect in this regard, either. John Anderson and Katie Parlante reviewed a number of cross-platform toolkits for Python known as GUI builders—sets of components for building windows, menus, buttons, and all the other onscreen "widgets" to bring a Python program to life onscreen for Windows, Macintosh, and Linux users. One of them, wxWidgets/wxPython, stood out (originally named wxWindows, the project dropped the Windows label to avoid a legal dispute with Microsoft). Wx wasn't quite there yet—the Macintosh tools were especially underdeveloped—but with a little more effort and time, Anderson believed it would work for OSAF.

As as alternative, the developers briefly considered using tools built by the Mozilla project, the open source Web browser built on Netscape's remains. Mozilla had GUI-building technology that worked well across all three operating system platforms and had a few years of development under its belt. But it wasn't as far along as wxWidgets, and Kapor's programmers, trying to peer one or two years out on the horizon and guess where to place their bets, worried that it might not develop as quickly as they needed. They also worried that programs built using Mozilla's tools wouldn't feel "native" to users of the different operating system platforms they planned to support; the menus and dialog boxes and windows might look odd and behave unexpectedly.

Kapor, Anderson, and their colleagues also briefly considered a more radical alternative to solving their cross-platform problem. What about the option of building their program entirely as a Web-based service, one that didn't need a separate client program but simply piggy-backed on the Web browser? Such programs had long been considered trifles or toys, but by 2002, software industry leaders were beginning to warm to their advantages. They meant that users could keep their data up on a server somewhere and call it down as needed to any computer they might be working on at home, at the office, or on the road.

Kapor knew all those arguments and considered them carefully, but in the end he felt that you just couldn't provide the kind of rich interaction his program would offer, the flexibility inherited from the "soul of Agenda," inside a Web browser. Not yet, anyway. There were too many limits on what you could do—on how easily you could manipulate and drag stuff on screen and how quickly data would update in a window.

Web-based programs just felt clunky, and clunky was the last word Kapor would ever want anyone to associate with his work.

■ ■ ■

With so many options available, a software project's choice of programming language often comes down to the arbitrary or the ineffable—a matter of taste or habit or gut sense. Programmer-essayist Paul Graham wrote that some coders favor Python simply because they like the way it looks—and that that's not such an unreasonable criterion: "When you program, you spend more time reading code than writing it. You push blobs of source code around the way a sculptor does blobs of clay. So a language that makes source code ugly is maddening to an exacting programmer, as clay full of lumps would be to a sculptor."

Although it was something of a leap of faith—you just weren't supposed to use a scripting language as the basis for an ambitious desktop software application!—Kapor, Anderson, and their team finally settled on Python. It felt right. And it was a bet on speed—the extra speed the language itself would lend OSAF programmers, and the additional speed Moore's Law would grant the machines of the eventual product's users, to make up for any sluggishness it might derive from Python itself.

Speed was on everyone's mind at OSAF, because progress through much of 2002 was halting. Hertzfeld, who had quit adding new features to the Vista prototype once the bigger architectural discussions kicked in, was beginning to sketch out what an address book or "contacts manager" for the new product might look like. Parlante had moved from user research into prototyping a calendar. Morgen Sagen had begun to create an automated build system for the project named Hardhat, a framework of code that would help the developers reassemble the program's building blocks after they made changes. Everyone was itching to start writing code.

But by the fall, OSAF was still in stealth mode. Kapor was a public figure; rumors had spread along the Silicon Valley grapevine that he was up to something, but what, exactly, remained a mystery. Kapor was loath to make a big-splash announcement without having a working product. That was classic "vaporware," and he'd been down that road with his company On Technology in the late eighties; it had the dubious honor of a place in the Vaporware Hall of Fame curated by a Macintosh columnist. On the other hand, if you intended to create open source software, you couldn't do it in secret; your project needed to be out in the open where interested volunteers could find it. When Kapor's friend Dan Gillmor, then a widely respected technology columnist for the *San Jose Mercury News*, pushed him to go public, he resisted at first and then decided it would be a good thing, a "forcing function" to get OSAF in gear and resolve some of the lingering uncertainties surrounding the project. Like the product's name.

The software industry has a long tradition of using code names for products under development. In the commercial end of the business, the programmers typically choose these placeholder names; later, the marketers pick a final, "shipping" name. In the open source world, the programmers usually have the first *and* last words, which is why so many open source products have names that are unpronounceable technical puns, abstruse references, or inside jokes. (Apache, the name of the world's most widely used Web server, is a pun based on how it grew in a piecemeal, or patchy, way. Then there are names like Mozilla, Knoppix, and Blojsom.) In general, the higher profile the project, the more care you have to take in picking a code name. Microsoft learned this in 2001 when Hailstorm, its working name for a system that would enable the stockpiling of troves of users' personal information, became a cudgel in the hands of critics.

Kapor didn't want to waste a lot of time wrangling over the code name for OSAF's project, and he managed to keep putting off the choice. Meanwhile, Morgen Sagen was building an online home for OSAF's work, a kind of nest for code called a CVS (for Concurrent Versions System) repository that allows developers to "check out" chunks of code to work on them, "check in" changes, and track everything that happens. Version tracking systems like CVS had made it possible for far-flung groups of programmers to work simultaneously on the same code base without stepping on one another's toes; they had made open source development possible. One day in September, Sagen went to Kapor and said that in order to set up OSAF's CVS repository, he needed a name for the file directories for the new project. Kapor took a few seconds to think and then said, "Call it Chandler."

"I like reading Chandler mysteries. I'd been rereading some," he later explained. Adopting one mystery author's name meant that others— Hammett? Christie?—could be drafted for future OSAF projects. A literary name, too, had a different sort of cachet than the usual techno-jargon. And naming the program for a classic mystery novelist would also playfully nod to the project's till-then surreptitious status.

Sagen worried that most people wouldn't think of the novelist but of the character from the television series *Friends* who had the same name. But he didn't push his objection. Chandler it would be.

■ ▌ ▌

On October 20, 2002, Gillmor's *Mercury News* column introduced Chandler to the world. OSAF simultaneously opened a Web site, and Kapor began writing a personal blog to chronicle the project. As Kapor had hoped, going public did force OSAF to specify more clearly what exactly Chandler was going to be, refining its vague aspirations into what Silicon Valley initiates called an "elevator pitch"—the pithy explanation you'd blurt out if you happened to luck into an elevator ride with some funder or pundit you wanted to win over. Chandler would be a cross-platform, open source, peer-to-peer personal information manager (or PIM) with a heavy focus on email and calendars—all infused with the dynamic spirit of Lotus Agenda. It was not exactly a simple sound bite; that elevator had better have some floors to travel. But it was enough to excite the software enthusiasts and open source devotees who noticed the announcement.

The word that Mitch Kapor was working on an innovative new "inter-personal information manager" hit the geek street with a jolt. Press coverage spread out in widening circles from Gillmor's initial piece, including a link from Slashdot, the Web's best trafficked and noisiest virtual hangout for programmers. Hundreds of outsiders posted their views and wishes to OSAF's newly public mailing lists. Most of the missives fell into these categories:

- ▸ I need Chandler immediately. Where is it? You mean you haven't written it yet?
- ▸ Why aren't you using [this or that particular] open source project as the basis for [this or that element] of Chandler?
- ▸ I will only consider using a PIM that provides this extremely specific feature that I have been using for many years and can't live without.
- ▸ Outlook blows. Microsoft sucks.

Slashdot's headline, "Mitch Kapor's Outlook-Killer," set the tone of the publicity: Kapor was taking on Gates. Chandler would slay Outlook. Up with open source; down with Microsoft!

The reaction was inevitable, but it pained Kapor. He knew that attacking Microsoft was asking for trouble—both because Microsoft's products are so deeply entrenched and because the company has a long record of squashing competitors when it perceives a threat. He thought he had been clear that his initial target market was not Outlook Exchange's big offices but, rather, "information-intensive" individual users and small companies. Chandler, he had said, aimed not to supplant the Outlook behemoth but to coexist with it. But he had underestimated the way his project would be framed by his own personal mythology. The slightest hint of a Kapor-versus-Gates grudge match inevitably enthralled the media.

Still, Kapor knew, Chandler would have to prove itself not by tilting at the Microsoft dragon but by winning users on its own merits. There was a long way to go, and there wasn't a whole lot of code yet, but the signs were propitious. Some very smart and experienced minds were working to solve problems that fascinated them. They stood on the tall shoulders of other open source projects. What could stop them?

Gillmor's October 20 column reported: "An early version of the calendar part of the software should be posted on the Web by the end of this year, and version 1.0 of the whole thing is slated for the end of 2003 or early 2004." Kapor wrote in his blog: "We will first put out code for developers to look at by the end of the year. It will be an extremely partial alpha. Optimistically, we could have a 1.0 by the end of 2003. Pessimistically, it will be 2004." But, he cautioned, this was OSAF's "original standard response"; he was "never entirely comfortable" with it, and "with Chandler in its current state, we don't yet know how fast we're going to be able to go, nor how far we have to get to the first release."

The guesses proved more than a tad optimistic.

CHAPTER 4

LEGO LAND

[NOVEMBER 2002–AUGUST 2003]

I n the roller-coaster cycle of the technology industry, the fall of 2002 represented a stomach-fluttering pause at the bottom of the dip. The dot-com boom had just about finished going bust. The parking lots of Silicon Valley's strip-mall-style office parks had become no-man's-lands. The post-9/11 economy had been holding its breath for a year, and the "America Is Open for Business" stickers plastered on downtown San Francisco storefronts seemed to beg the retort: Yeah, but who's buying?

This tech industry ice age had left legions of programmers un- or underemployed. A good number of them had turned to open source projects as a way to keep their skills honed and their names in circulation while scratching whatever programming itches they might have. This surplus idleness fathered a significant amount of new software, from little side efforts to ambitious projects that would grow, when the industry wheel turned once more, into a new generation of start-up companies.

OSAF certainly benefited from this timing: It was a bright light on a dark plain, and it beckoned to programmers looking for projects. Mitch Kapor had always figured that the organization would depend on both paid employees and altruistic volunteers, and in Andy Hertzfeld it already had one high-profile example of the latter. The flurry of OSAF news coverage that followed Gillmor's column filled Kapor's inbox with encouragement and offers of help. Steve Jobs called Kapor—it was the two men's first conversation in a decade—inviting the OSAF team down to Apple to talk about how they might collaborate.

Kapor took special note when Lou Montulli's name turned up in the signature of an email to OSAF's just-opened "dev" mailing list—a message suggesting that OSAF look again at Mozilla's toolkit for building its user interface. Montulli and his pal Aleks Totic had seen the Slashdot article about OSAF; soon after, they got in touch with Kapor to talk about volunteering on Chandler. Both young men were famous in Web circles: Totic had been part of the group at the University of Illinois, Urbana-Champaign, that built Mosaic, the first popular graphic Web browser (he had written the Macintosh version), and then decamped with his colleagues for Silicon Valley as part of the founding team at Netscape in 1994. Montulli had written a very early text-only Web browser while a student at the University of Kansas and had also been recruited to join Netscape's start-up crew. He is probably best known as the inventor of the "cookie"—the much-maligned and oft-misunderstood file that Web sites plant with a browser to track visitors and their data from one session to the next. When Netscape launched the Internet boom in August 1995, with a skyrocketing initial public offering, Montulli and Totic, like all the Netscape founders, made their fortunes. Good friends who liked working together, they both left Netscape as its end neared and helped found Epinions, an ambitious e-commerce site; they then left that company when it got sucked into the downturn's maelstrom.

In 2002, they found themselves at loose ends. Lou was mostly living at Lake Tahoe and skiing. Aleks was between projects, too. Chandler sounded cool. Maybe Kapor wanted some help?

Kapor knew that, like Hertzfeld and John Anderson, he was an "old applications guy," as he put it. He had always worked with programs that computer users ran individually on their own computers. But Chandler was going to be different: Sharing data over a network was central to its promise. Montulli and Totic offered the perspective of programmers who had spent nearly a decade creating software that depended on the Internet for its critical functions. That, Kapor thought, was likely to come in handy.

Totic and Montulli joined OSAF as volunteers around the same time that Kapor began to hire for other jobs, filling out the organization's roster as it moved into the spotlight. The team was heavy with Apple alumni, who perhaps found in Kapor's vision an echo of Steve Jobs's change-the-world fervor. Taking on the job of Chandler's product manager was Chi-Chao Lam, a veteran but still boyish Silicon Valley entrepreneur who had worked at Kaleida Labs, the ill-fated mid-1990s joint venture between IBM and Apple, then founded a pioneering Web-based advertising network. Pieter Hartsook, a longtime Macintosh trade journalist and industry analyst, signed on to handle OSAF's marketing and PR. And two more programmers joined the coding crew: Jed Burgess, a recent Stanford computer science graduate, would work closely with John Anderson on the GUI. David McCusker, a programmer who had done stints at Apple and Netscape, was hired to work on Chandler's approach to storing data.

In another version of software's layers of abstraction, software developers often talk about "front ends" and "back ends." You, the user, are in front, and the front end is the part of the program that deals with you—that offers windows and dialogues and mouse pointers, tells you what's happening, and provides ways for you to input and output information. The back end is where the results of front-end events and inputs go so that the computer can make sense of them, save them, and retrieve them. (In Chandler's prototype, Vista had been the front end, and Shimmer the back.) Front ends are supposed to be elegant, intuitive, versatile; back ends are supposed to be invisible, efficient, rock-solid. The front end talks to people; the back end talks to bits. In *Star Wars* terms, the front end is the butlerish C3PO; the back end is the unintelligible R2D2.

Kapor had a relatively clear vision of what he wanted Chandler's front end to do and how it would embody the soul of Agenda. Whatever form Chandler's back end took would have to support that vision. But exactly what form would that be? As OSAF's programmers—in particular Katie Parlante, whose calendar was the farthest along of Chandler's pieces—struggled to begin writing the program, they kept bumping up against unresolved questions: Would the Chandler repository, the program's storehouse of data, save everything in RDF "triples" the way Sagen's Shimmer prototype had? How would the programmers save the Python objects they were constructing Chandler with? Couldn't the repository store those, too? If Chandler was going to support peer-to-peer data sharing, how would the repository manage that? What if your Chandler repository was stored on a computer at home and you wanted to use it at work?

What was the plan? The programmers understood that until they had answers, no significant work could proceed. They were all stuck, stymied—in the software business's preferred term, *blocked*.

The back-end work Chandler needed led directly into a maze of tough technical decisions, and McCusker quickly found himself at the center of it. An enthusiastic blogger who used his Web site, named Treedragon, to chronicle his own side projects developing new programming languages, McCusker filled blog entries in the days after his start at OSAF with enthusiastic, point-by-point responses to technical emails. One email he posted from a developer whom he didn't name read:

> Please don't blow it and reinvent what Python and
> Zope [a Python-based project] developers have
> already created. Python and Zope have been in
> development for years and there is a treasure
> trove of great technology in this tool chest. The
> key to making Chandler successful is in reusing,
> rather than reinventing this work. Integrate the

```
prior successes instead of reinventing them. See
how little code you need to write in creating
Chandler . . .
```

McCusker replied, "We want to reuse existing code as much as possible rather than write new code. Making progress quickly involves being very reticent about forging out into new coding areas."

But it was one thing to express a general wish to reuse code, and quite another one to commit to specific choices of actual pieces of existing code. Each alternative had attractions and drawbacks; each approach had advocates and critics among the growing circle of attendees at OSAF's meetings. No one wanted to make the wrong choice. In meetings through November and December, culminating in a marathon right after the New Year, the Chandler team struggled toward an elusive consensus.

The RDF-based Shimmer repository was something Morgen Sagen had built expressly as a prototype; it couldn't simply be hitched onto the real Chandler. Besides, John Anderson had never gotten the RDF religion. The whole RDF enterprise had a reputation for academic complexity and impracticality. There were lots of papers about the Semantic Web, but not a lot of working software. As one programmer after another had a look at the world of RDF, each came to a similar conclusion: It was "scary."

Anderson knew how much work programming Chandler's user interface would be. He had been there before and understood how critical it was to keep that job manageable; it was the area most likely to cause endless delay. His chief requirement for the repository was to make things easier for the front-end developers. And he knew just what they needed: a system for "object persistence." Rather than store Chandler's data in the RDF subject-verb-object "triple" format, Anderson wanted the program's data and Python code stored together as objects (in the "object-oriented programming" sense of the term) that Chandler's coders could easily grab, manipulate, and save again.

The most common kind of database on today's computers—such as Oracle's hugely successful product and the increasingly popular open source MySQL—is "relational": It stores information in vast tables that break data up into discrete units and then allow you to construct complex queries to combine it in useful ways—queries like "Show me all the entries from this date whose last names begin with 'c' but excluding those missing the street address." Relational databases also typically provide great reliability by using a "transactions" model that logs every single change to every iota of data so you can always reconstruct your information in the event of a crash or data corruption.

The easiest way to get "object persistence" was to use a different kind of database, an object database, one that was designed not just to store data but also to keep track of the relationships between data and the code associated with the data. This approach could make life easier for programmers; it also tended to make for slower and sometimes less reliable databases. As it happened, the community of Python programmers had already created an ambitious and well-known example of an object database. It was called ZODB, for Zope Object Database. McCusker's anonymous email correspondent had referred to it. At that time Python creator Guido van Rossum worked for the company that produced it.

ZODB was the answer to Chandler's problems, Anderson concluded. At first Sagen, who had written Shimmer and had been in charge of Chandler's back end, didn't agree, but he admitted that his experience was more in programming for servers than for clients—the programs that actual users employ—and eventually bowed to Anderson's choice.

Meanwhile, however, Lou Montulli and Aleks Totic had begun attending OSAF's meetings and offering observations from an entirely different direction. If Chandler was going to fulfill its promise of easy peer-to-peer sharing, Montulli argued, it had to be built for speed across a network: You had to be able to hand stuff off quickly "over the wire"—programming shorthand for passage across an open network like the Internet. He laid out a vision for Chandler that separated the client and its user interface from the repository. Each Chandler client would access a repository using an Internet-style protocol; he named it RAP, for Repository Access Protocol. RAP was how Chandler would talk to the repository even if your client and repository both sat on the same computer on your desk. With that one big architectural choice, Montulli maintained, you'd resolve all the questions about remote access to your data, and you'd make sharing data easy.

The questions surrounding remote access and sharing had become more urgent for Kapor because, in the flurry of enthusiasm for Chandler that followed its announcement in October, a working group of university chief technical officers contacted him. The committee, the Common Solutions Group, or CSG, represented two dozen institutions, including Berkeley, MIT, Harvard, Yale, and Stanford. CSG faced the problem of providing campuses with a cheap, open, reliable software platform for calendars and email. Chandler, some of its members thought, might be just the thing. It was an intriguing opportunity for OSAF; it also meant that Chandler's programmers now had to consider a whole new set of use cases. The original plan was to target Chandler at "early adopter" individuals and small businesses that were "info-centric"—whose work involved dealing with a large volume of email, appointments, and other random data. Now the developers were also thinking about students and faculty spread across a campus—big institutions with diverse populations of mobile users.

Anderson wasn't enthusiastic about Montulli's RAP idea, but he was willing to go along as long as it didn't get in the way of his top concern: object persistence. Sagen ducked out of the debate completely. "I was feeling a little unsure of myself," he later explained. Surrounded by programmers with imposing pedigrees, he would think, "What do I know?"

McCusker sat in the middle, trying to figure out how to keep everyone happy as the demands on his part of the project multiplied. He was an experienced juggler; for several years his home page featured a picture of him juggling at the beach—five rainbow-colored balls arcing from one hand to the other. But his dexterity was being taxed by the lengthening list of conflicting requirements that Chandler's designers were compiling for him. "What the OSAF folks wanted—the entire laundry list of desires—was impossible," he later told me.

It would have been a tough position for any programmer, even one more bullheaded than McCusker, who enjoyed looking at problems from multiple perspectives. But it was also clear that he was having a hard time pulling a plan together because his life outside OSAF had begun to fall apart almost immediately after he joined the Chandler team. His marriage broke up, he grew depressed, and he came down with a miserable case of repetitive stress injury that forced him off the keyboard and onto a speech-recognition dictation system. Over the holidays he decided to change his first name from David to Rys, and soon he began to change his appearance, too: He let his hair grow long and dyed it blond. He began wearing lipstick. He blogged about his interest in an "alternative lifestyle," though he warned his readers not to "speculate incorrectly": "You almost certainly are getting something wrong. However, that's not my problem, and I'm not particularly upset by being misunderstood. In addition, I happen to think ambiguity is funny, and I enjoy confusing you." He was clearly going through an identity crisis of sorts, and wherever it was taking him, it wasn't helping him solve Chandler's repository problems.

OSAF's environment was low-key and tolerant, and the programmers quickly adapted, training themselves to use McCusker's new moniker. Mostly, it seemed, they didn't care about how he looked or what he wanted to be called, but they did want to get some decisions made about the repository so that they could move forward with the rest of the program. When OSAF had publicly introduced itself in October 2002, Kapor and his team got in the habit of talking about the "January release" of a first draft of Chandler. By early January, as it became obvious they were nowhere near any kind of release, they switched to talking about an undated "first public release" of code.

It's a rainy Thursday morning early in January 2003, and the developers have gathered at the Belmont office. They've already filled up the white-board in the main conference room, so they move into its antechamber, a space piled high with extra furniture. There aren't enough chairs, but two big blue plastic storage bins, flipped on their sides, offer bar-stool-style perches. Montulli, sporting a few days' growth of beard, sits on one at the back; next to him, Totic buries his head in a laptop. Anderson, graying but ruddy-faced, pulls a chair into the doorway. Hertzfeld pulls his feet up onto his seat and crosses his legs.

They dive into the problem of ensuring that Chandler-across-a-network will not be a laggard. Anderson describes a design that would "pre-fetch" a screenful of objects, which means that if you had an email inbox on your screen with twenty-five messages, the program would be ready to show the next twenty-five if you scrolled down the page.

Montulli suggests that it would be better to start streaming as much data as possible as soon as the user asks for something—more like a Web browser. "The most popular thing we did at Netscape," he said, "was to have the page render on the screen even before everything had come in." That simple choice made the Web feel faster and more responsive at a time when most people were accessing it via slow modems and cursing it as the World Wide Wait. Chandler, Montulli argues, could learn from that experience.

As the argument proceeds, though, it becomes evident that he is really advocating something more basic. He never comes right out and says, "Let's dump ZODB," but he implies that Chandler doesn't need it, that it will only get in the way.

Anderson feels that Montulli is going too far. "Let's go back to square one," he says. "Do we use object-oriented programming?"

"Yes," Montulli answers.

"Do we use Python?"

Montulli nods. Totic, who has mostly kept quiet, looks up from his screen and weighs in. It's just not clear that ZODB is exactly what Chandler needs, he explains in diffident tones, gently accented from his native Belgrade. ZODB, it turns out, is really just a *layer* for object persistence that sits on top of a lower level open source database called Berkeley DB. In the time it would take OSAF's programmers to study ZODB, figure out whether it suited their purposes, and tinker with it to get it to fit, Totic figures, they could build their own object persistence code directly on top of Berkeley DB.

Anderson's "I've found the perfect solution" versus Totic's "We could do it ourselves just as easily": Here, once more, was the archetypal dilemma of software reuse. Build or borrow? Virtually every software project sooner or later arrives at this fork in the road. (The full phrase traditionally is "Build, buy, or borrow?" From a technical perspective, though, "buy" and "borrow" are similar, the commercial and open source sides of the same coin.) The world is full of code that someone else has written. Shouldn't it be easy to grab it and use it for your new project? Why then do so many programmers glance at existing code and declare authoritatively that they could do it themselves, faster, easier, and better?

■ ▍ ▮

When people dream of streamlining the work of making software, most often they dream of standardized plug-in parts. James Noble and Robert Biddle, two scholars in New Zealand who sometimes write together under the sobriquet The Postmodern Programmers, dub this vision the Lego Hypothesis: "In the future, programs will be built out of reusable parts. Software parts will be available worldwide. Software engineering will be set free from the mundane necessity of programming." Pull some pieces off the shelf, snap them together, and presto—working software, with no painful coding!

Programmers who have tried to travel the road to this utopian vision of programming have almost always found it blocked. Noble's and Biddle's research identified one of the most significant obstacles. They and two colleagues examined a broad sample of software objects used in real programs built with object-oriented techniques and found that these building blocks weren't Lego-like in any way. If software components were like Lego bricks, they'd be small, indivisible, and substitutable; they'd be more similar to one another than different; they'd be "coupled to only a few, neighboring components." But Noble and Biddle found that the components in actual programs varied enormously in size, in function, and in their number of connections to other components. They were "scale-free, like fractals, and unlike Lego bricks." When they peered under the hood of real programs, Noble and Biddle observed what they called "pervasive heterogeneity": Everywhere you looked, the only constant was that nothing was constant. Imagine a Lego set in which some parts are a half-inch long and others are a half-mile long; some are made out of hard plastic and others are liquid or gas; some click together with the familiar rows of protrusions, and others connect via glue or solder or string.

No wonder it hasn't been easy to produce a Universal Snap-In Software Kit. Yet despite repeated failures, the Lego dream has cast a long shadow across the story of modern programming. And for all the obstacles, each new generation of programmers does find better ways to borrow instead of build.

The concept of reusable code dates back to software's antediluvian roots, when Maurice Wilkes (the British programming pioneer who had the glum realization about his lifetime of debugging) and his colleagues in postwar Britain defined the subroutine: a chunk of code that could be invoked from anywhere in a program to accomplish a certain task and that, having finished its assignment, would hand the reins back to whichever part of the program had called it. Wilkes almost immediately began collecting subroutines into libraries—"in order that every user should not have to start from the bottom." From that day on, most programmers have also worn the hat of code librarian.

But these libraries of code routines typically remain islands. True standardization has eluded the field; in software there are dozens of competing standards everywhere you look. Each point of differentiation in a computer system—what central processor chip are you using? what version of which operating system? what programming language? what data format? and so on—offers another point of failure for the snap-it-in Lego dream. As Robert Glass, the author of a number of books on software engineering, puts it, programmers long ago solved the problem of "reuse-in-the-small," building libraries of subroutines to automate the drudgery out of their work. Still unsolved is the problem of "reuse-in-the-large," making big reusable software components that are actually useful and used. "It is not a will problem," Glass wrote. "It is not even a skill problem. It is simply a problem too hard to be solved, one rooted in software diversity."

In the 1980s, a computer scientist named Brad Cox became obsessed with the problem of constructing programs out of reusable parts. He had designed an extension of the C programming language, called Objective-C, intended to streamline the production of reusable software objects. But when it didn't catch on, he began to view the issue from an economic rather than a technical perspective. In 1990, Cox outlined a new approach to the building and funding of software in which programmers would code and sell small components for others to assemble into working systems.

We are still building our software cottage-industry-style today, Cox argued. We need to "transform programming from a solitary cut-to-fit craft into an organizational enterprise like manufacturing." Today's software developer is like the colonial-era gunsmith, lovingly handcrafting every nut and screw. We haven't yet made an industrial revolution leap from "filing away at software like gunsmiths at iron bars" to "commercially robust repositories of trusted, stable components whose properties can be understood and tabulated in standard catalogs, like the handbooks of other mature engineering domains." The computer hardware industry, despite the enormous complexity of the technology it produces, had managed this transition: Users with screwdriver-level skills could plug in cards—circuit boards to extend a computer's functions—without having to understand anything about the chips on the cards. Similarly, manufacturers can build those cards from off-the-shelf chips without worrying about how the chip maker produced them. Why can't we build software the same way?

Cox tried. He founded a company called Stepstone aimed at "providing pluggable chip-level software components to C system builders." But despite some small successes, the approach didn't take off. He later wrote, "The bad news is that this experiment has shown that it is exceedingly difficult, even with state-of-the-art technologies, to design and build components that are both useful and genuinely reusable, to document them so customers can understand them, to port them to an unceasing torrent of new hardware platforms, to ensure that recent enhancements or ports haven't violated some existing interface, and to market them to a culture whose value system, like the Williamsburg gunsmith, encourages building everything from first principles."

The rise of the Internet led Cox to take another swipe at the problem. In his 1995 book, *Superdistribution*, he outlined a scheme for using the Net to build a new kind of market for software components. In computing today, he pointed out, it is nearly impossible for a piece of code to track when it is being copied, but it is easy for it to track when it is being used. So why not write software that can be freely copied but that tracks each use and charges accordingly? Say I'm a developer building a new email program. In the course of my construction I find a piece of code somebody wrote that filters spam, and I decide to incorporate it. You get a copy of my software for free but get charged each month to use it, and when I get your fee, I kick a portion of it over to the author of the spam filter.

Cox hoped that "superdistribution" would create incentives for the evolution of a bustling automated market for reliable software components, but once more, his ideas did not catch on. Today's Web-based software services—where the user never downloads, touches, or "owns" the running software but instead just accesses it through a Web browser for free or for a modest fee—come closest to the world of superdistribution and metered use of software components that Cox imagined. Meanwhile, programmers, instead of installing hunks of code that know when they're being used and charge for it, have gravitated toward the open source free-for-all.

Cox, now a consultant living in Manassas, Virginia, says that although he uses open source software daily and considers himself a Linux hacker, he doesn't see the open source approach as a realistic answer to the problem of software reuse. "They do have an economic model that works for people that are motivated by reputation more than money," he says. "But most folks are motivated by how to pay the mortgage. And I'm more interested in how to mobilize that majority to make reuse work."

Beyond the technical hurdles and economic disconnects that bar software developers from their promised Lego Land, simple human nature also sometimes stands in the dream's way. Sure, great programmers rewrite and reuse, as Eric Raymond wrote. But programmers have always felt a certain pride of authorship, and telling them to just pull parts off the shelf can rub them the wrong way.

Larry Constantine, author of a popular column in the 1990s called "The Peopleware Papers," offered one pragmatic explanation for why programmers did not flock to Cox's ideas. "Unfortunately," Constantine wrote, "most programmers like to program. Some of them would rather program than eat or bathe. Most of them would much rather cut code than chase documentation or search catalogs or try to figure out some other stupid programmer's idiotic work. . . . Other things being equal, programmers design and build from scratch rather than recycle." Sometimes software developers fall prey to the "not invented here" syndrome, in which confidence in your own skills and your organization's prowess leads to distrust of anything created by anyone else; sometimes they just don't feel like taking the time to study someone else's work or even to look around to see if anyone else's work might suit their needs. "If it takes the typical programmer more than two minutes and twenty-seven seconds to find something," Constantine wrote, "they will conclude it does not exist and therefore will reinvent it."

The twin revolutions of open source development and the Internet have certainly begun to change that habit. Google has shortened the process of finding many things to a duration that even the programmer's two-minute-and-twenty-seven-second attention span can handle. And languages like Perl and Python offer programmers vast troves of code for accomplishing mundane but time-consumingly complex tasks, like processing dates and times or sorting pieces of text. For instance, the Comprehensive Perl Archive Network, or CPAN, hosts (as I write this) 8,107 "modules" by 4,359 authors; each is available for free on the Net.

These storehouses of program components are probably the closest the software world has come to the Lego dream. And yet they remain less used and less useful than their most ardent contributors might wish. There are often dozens of different choices and different versions of those choices in the same category. The volume of available code and the number of things each package of code can do are so great that even veterans sometimes forget what's available; they end up hunting in vain for what they need or going off and writing some function from scratch that was already available in ready-to-use form.

When Python expert Anthony Baxter visited OSAF, he gave a talk about building an Internet telephone application named Shtoom. Handling audio requires good performance from your code; it's "real time"—you have to keep up with the sound as the data comes in or the audio will break up, so you have to process what you get fast. "Raw Python could almost handle it," Baxter said, but it needed just a bit more streamlining or optimization. Baxter told of hunting around for different solutions to the problem: a special Python compiler called Psyco, a math processor called Numeric, a tool called Pyrex that compiled Python into C code.

Then he remembered: Python itself offered a module called "audioop" for efficiently manipulating audio data. The standard distribution of the programming language already did exactly what he needed! He of all people should have known that: He was the release manager for the most recent version of Python, the programmer in charge of shepherding the program language and its packages through development and testing to final release. And yet he had completely forgotten about the existence of the tool he sought.

The batteries were indeed included, as Python devotees like to say. But with so *many* batteries, who could keep track of them all? The problem extends to every language and across every realm of software building. "Keeping up with what's available in the libraries," says programming expert Ward Cunningham, "is the number one information overload challenge."

Sourceforge, an open source warehouse on the Web, hosts the code for more than 100,000 projects. Few have a large population of users; a good number of them are simply one-person operations. There are programs here for accomplishing just about anything you might want to do with a computer—from processing home videos to running a business's books. But it seems to be in the nature of programmers in general, and open source programmers in particular, to want to reinvent the wheel. So if you are looking to find a video conversion program on Sourceforge, you have 190 to choose from. This penchant among programmers is deep-rooted; the phrase "Yet

Another" prefixed to a program's name long ago became a way of acknowl-edging that whatever you were doing, many others had already done it before. There's YABB, "Yet Another Bulletin Board," and YASSP, "Yet Another Solaris Security Package," and Yet Another Bittorrent Client, and on and on. When two Stanford grad students started up Yahoo! in 1994, the name was a smart-alecky acronym for Yet Another Hierarchical Officious Oracle.

If Mitch Kapor was sure of anything about Chandler, it was that he didn't want the project to be "Yet Another"—not yet another Outlook clone, not yet another garden-variety personal information manager, not yet another anything. The whole point of Chandler was to build something truly dif-ferent, software that would bend to your needs. Programs like Outlook seg-regated emails from tasks and tasks from appointments and appointments from notes; they might give you a little toolbar with icons that allowed you to view each type of data, but you couldn't mix and match, you couldn't relate one to another, you couldn't store one in more than one place, and you couldn't effortlessly transform one into another. In Chandler, every-thing that you stored—email and addresses and events (appointments) and to-dos and random notes—was simply an "item," and items would be organized and displayed however you damn well pleased.

Conventional programs segregated information of various types in what OSAF's developers began to call "silos." The Chandler developers' battle cry was: Level the silos!

It was not the first time such a cry had been raised. Digital-age maverick Ted Nelson had propounded the idea of computers as "dream machines" and engines of personal liberation in the 1970s and invented the term *hypertext* to describe writing with embedded links that let you jump from one place to another. (Nelson views today's Web as a bastardization of his more complex vision.) Nelson also coined the word *intertwingularity* as a label for the kind of complexity that informational silos ignore: "People keep pretending they can make things deeply hierarchical, categorizable and sequential when they can't," he said. "Everything is deeply intertwingled."

For all the contempt heaped upon them, however, silos—discrete containers that hold a particular set of similar items in one and only one place—have stubbornly persisted. The graphical interface of modern computing, the GUI, had evolved around physical-world metaphors—desktops, files, and folders. And in the physical world a file could exist in only one place at a time. In the virtual world of computer data, there was theoretically no reason for this limitation. Of course, somewhere down near the bottom of the layers of abstraction, in the physical sectors on your computer's disk drive, the file really does exist in a single place. But build enough layers of software on top of that reality, and you could *present* that file to users in multiple ways, allowing them to access it from as many different points as they wish. They might appreciate that.

As any everyday office worker knows, however, you don't often encounter such freedom in today's computing. Silos are everywhere. For one thing, there's that preference our brains have, shaped by their evolution in the physical world, for thinking of items as residing in a single place. And because a piece of data really does exist in a single place on the computer's bottommost layers, it's far easier to build software on that basis than to try to do something different. Less can go wrong. The programmers' "mental model"—the picture they have in mind of how the program's parts fit together, of where the data is and what should happen when the user looks for it or moves it—aligns with the users'.

Silos are common because they are simple, reliable, and unambiguous. They also reflect the inherent binary nature of digital computing, in which a bit is set at either one or zero, on or off, true or false. It is not impossible to build ambiguity into such a system, to design a structure that lets you put one thing in more than one place. But you have to work a lot harder—as the Chandler team was finding.

Chandler's repository had to support the kind of flexibility and fluidity and silo-less-ness that represented, in Kapor's mind and to everyone else at OSAF, the soul of Agenda. It also had to support the kind of over-the-network synchronization that Chandler's peer-to-peer design demanded. Would ZODB be adequate? No one could say for sure. It was awfully hard to say such a thing with any confidence about a piece of code unless you had written it yourself.

Here is one of the paradoxes of the reusable software dream that programmers keep rediscovering: There is almost always *something* you can pull off the shelf that will satisfy many of your needs. But usually the parts of what you need done that your off-the-shelf code won't handle are the very parts that make your new project different, unique, innovative—and they're why you're building it in the first place.

■ ▮ ▪

The January meeting goes on for several hours, circling the question of whether to use ZODB but never resolving it. After a lunch break, Kapor stands up at the whiteboard, bouncing on the heels of his running shoes, his white shirt beginning to climb out of his trousers. He tries to precipitate something from the discussion.

"I've been thinking about why this is so hard," he begins, and starts listing reasons on the board. He asks whether Anderson is focused too much on making it easy for the programmers at the expense of the users. He starts outlining requirements for the repository:

(1)   It has to make life easy for the Python programmers.

(2)   It has to operate efficiently over a network.

(3)   It needs to be able to handle very large individual data items and very large numbers of items.

(4)   It has to be reliable, using database "transactions."

(5)   It has to support searching and indexing.

Kapor adds several more items to the list, then goes back to the top and writes above the first line:

(0)   User experience.

At the day's end there's no concrete plan for the repository, but everyone has action items. John Anderson will hold a meeting with Lou Montulli and Aleks Totic and try to give them a crash course in object-oriented graphic interface (GUI) programming and why he believes "object persistence" is so critical. Katie Parlante will prepare a document outlining what approach to data would make life easier for the GUI programmers. And Rys McCusker will study ZODB and see how well it can support Kapor's list of requirements.

"I note," Kapor says, "that these new tasks completely diverge from what we now have on our development schedules. I'm confident that they'll converge again somewhere down the line."

■ ▮ ▮

Through the winter of 2003, the developers kept taking tentative stabs at commencing the construction of the real Chandler—trying to leave the world of prototypes and start producing code that would form the basis, however embryonic, of the final program. Anderson mocked up the first crude version of a user interface in wxWidgets and wxPython. Hertzfeld whipped up a quick rendition of an address book or contacts manager. But without knowing how the program would store its data, nobody wanted to write too much; it felt like throwing walls up before you had even picked a site for a foundation. A functioning program still seemed like a mirage: No matter how many meetings the developers held or how many decisions they thought they had made, the goal moved no closer.

Part of the problem was that although Kapor was the ultimate arbiter of all things Chandler, he felt out of his depth in the repository debate and didn't want to dictate one approach over another. Part of the problem was also that Anderson, though strong and clear in his technical preferences, was neither confrontational nor impulsive, and when people raised questions and concerns, he took time to consider them.

But mostly, everyone was waiting for Rys McCusker to start building the repository. McCusker, meanwhile, felt that Chandler had gotten derailed by Montulli and Totic, who tended to discount decisions the project had already made, like the choice of Python and ZODB. If he hadn't been going through such personal turmoil, he reflected later, he might have mounted stronger technical arguments against them; instead, he watched in dismay at what he viewed as their obstructionist tactics. For their part, Montulli and Totic thought McCusker wasn't up to the job of cranking out something that OSAF could use quickly. And Montulli questioned whether the whole object-persistence approach—which he viewed as more of an ivory-tower "pipe dream" than a field-proven programming technique—was worth the effort.

On February 12, McCusker wrote in his blog: "I have not yet written any code for Chandler. I have written no Python code. I have written no C++ code. I have not checked anything at all into the source code control system."

At the end of January, as the sense of stasis grew stronger, Hertzfeld had suggested that—repository or no repository, and even if they weren't ready to show their code to the public—it was time for the developers to start assembling the code they had each been separately working on into milestone releases. The programmers began regularly depositing their code into OSAF's CVS repository, the source-control system that tracked every change to the code and allowed "rollbacks" to any previous version.

And so, on February 20, 2003, OSAF released the very first Chandler milestone. *Release* is not quite right, though; it was an internal release only, just for the OSAF developers, not to be made public. It was fenced around with every possible disclaimer. It was "pre-alpha." It continued to rely on Sagen's Shimmer prototype for the back end. It looked a little bit like a Web browser, with back and forward buttons. It offered a rudimentary calendar and contacts manager, because Katie Parlante and Andy Hertzfeld had coded them.

It didn't do much of anything. But it was something, at least, and after nearly four months since the media trumpets had first blown for Chandler, that gave everyone at OSAF some relief and a lift.

On the day the programmers released this version of Chandler to themselves, I bumped into Kapor by the water fountain. "We're doing it!" he said with a smile. "We're actually developing software!"

■ ▎ ▏

Months earlier, when everyone thought progress would be swifter, Kapor had agreed to deliver a presentation about Chandler in April at the O'Reilly Emerging Technology Conference, an industry event focused on peer-to-peer software and other new technologies. Now it was clear there was no chance of finishing Chandler's new repository in time for that event. But the conference could still serve as a useful deadline, helping to push the developers into gear for the long-brewing initial public release.

There was already more code on tap: In addition to Parlante's calendar and Hertzfeld's address book, Chao Lam, the product manager, had whipped up in his spare time a Chandler "parcel," or add-on program, for reading feeds of blog posts, known as RSS feeds. He called it Zaobao after the Chinese word for newspaper. Zaobao offered the first glimpse of Chandler's "extensibility"—the idea that the program would go beyond the traditional PIM world of email, calendar, tasks, and addresses, and allow programmers to write extensions so that users could organize all sorts of information in one silo-less program. What if you read a blog post about an event in Zaobao and wanted to send it to a friend or turn it into an entry on your calendar? You'd be able to do it right in Chandler. No cut-and-paste, no switching from one program to another.

Cool! But still very, very far off. There were a few other novel features in the works: Even without a new repository, Hertzfeld had found a way to rig some rudimentary sharing between Chandler users by adopting an open source instant-messaging protocol called Jabber. But mostly, Kapor and the rest of the Chandler crew saw, the first version of Chandler that the public would glimpse was not going to be too impressive. That was common enough in the open source world, where the norm was to put code out as soon as you had any and keep working on it in full public view with the aid of any volunteers you might attract. But OSAF had put an ambitious stake in the ground when it first announced Chandler; now, six months later, the developers feared they were going to let down the throngs of mailing list onlookers who had sent in encouraging messages and suggestions.

At a meeting late in March, Michael Toy, who had recently joined OSAF to manage the programming team, declared, "The goal of this release is to demonstrate that there is life inside OSAF."

"*Intelligent* life," Andy Hertzfeld interjected.

"Right," Toy responded. "It does not need to be useful as long as it's living and showing signs of growing."

"And intelligence," Hertzfeld repeated.

■ ■ ■

Chandler 0.1, as the first public release was labeled, went live on OSAF's servers on April 21, 2003, at 3:07 P.M. Pacific time. Within twenty-four hours it had been downloaded fifteen thousand times.

It had evolved a bit from the first internal release. You could add appointments to the calendar, and names and addresses to the contact list. But it remained crude, slow, and not very usable for anything. When you started it up, the opening screen declared: "Please understand that our underlying database is undergoing massive changes at the moment. What we're using now *will not be* what we use in the future. This means that"— here the text shifts to red—"you should not store important information in Chandler yet."

Kapor wrote on his blog: "I've been referring to Chandler 0.1 as our 'ultrasound' release: the fetus won't be viable outside the womb, but if you look closely you can see tiny arms and legs waving around and you can believe it's going to turn into a real baby eventually."

■ ■ ■

Around the same time as the 0.1 release, Kapor called Rys McCusker into his office and told him that things weren't working out. Chandler needed a repository, and he just wasn't getting it done.

In the previous month, actually, McCusker had drawn up a final repository design that seemed to cut the difference between the alternative approaches that Anderson and Montulli had advocated. The lead developer for ZODB had visited OSAF for a couple of days' consulting. There were some lingering questions about how ZODB would work with Montulli's

Repository Access Protocol, some qualms about ZODB's reliability and efficiency, and some concerns that its code was hard for outsiders to grasp. But at least there was a plan.

Still, there was little progress toward actually building a working repository. Given everything that had happened in McCusker's life over the previous months, that was perhaps not surprising. You could almost think of his as a disability case, Kapor told the programmer. Whatever the cause, things just weren't working out.

McCusker went home and later posted on his blog: "I don't understand why I am sometimes described as a database expert, despite the fact I disclaim such a description periodically. But I've noticed my replacement tends to be described as a database guru, and I had no idea that's what I was supposed to be. I do a lot of plumbing in general, which involves moving bytes around inside memory and between memory and secondary storage, which typically has the character of a database. But an expert on databases I'm not."

An expert on databases was what OSAF now began seeking.

■ ■ ■

It's May 21, 2003, and as part of the weeklong series of meetings aimed at taming Chandler's "snakes," problems that seemed beyond solution, the developers gather once more in front of the Belmont whiteboard to talk about the repository. Michael Toy presides. There's one new face at the table: Andi Vajda, who started the day before as OSAF's repository developer.

"This has the potential for being the longest or the shortest of these meetings," Toy begins. "I believe we have the following goals for our Chandler data storage solution." He sketches an outline on the board:

(1) Provide programmers with as revolutionary a data model as users.

(2) Data can live anywhere.

(3) Data is safe from corruption.

(4) Data is quick to get.

(5) Data can be large.

Montulli objects to the word *revolutionary*; he thinks it reeks of hype. Hertzfeld points out that the way the first point is phrased, you could get away with providing both groups with a mediocre solution and still "meet" the requirement.

Kapor gets a little impatient: "That's logic-chopping."

"We got here by having high goals," Toy continues. "Part of our struggle is in not valuing the goals equally. But there's no disagreement over the list itself. The biggest fear is that the goals are simply not meetable. We do have a proposed design. It's got some holes. It's time to talk about those holes and make sure we have plans to fill those holes so that never again, as long as I live, do we roll our eyes when the word *data* comes out of our mouths. This is *the* snake that must be killed."

"And in this meeting," Hertzfeld says, "we kill the snake—we don't just make plans to kill the snake."

"Yes," Toy answers.

"The dragon has many swords in it."

"There's also some bodies of brave knights who walked into the flames."

Kapor cuts in. "Look, we already have a provisional plan based on Rys's work that will meet our requirements. Andi, having just got here yesterday, is going to want some time before he signs in blood on the dotted line. But we shouldn't be starting from first principles all over again."

The programmers begin to discuss the series of issues that had always hovered over the repository problem: How would you make a repository work efficiently across the Net—say, when you were trying to load a large file like a song in MP3 format? How would you keep repositories secure? How would you share between repositories or keep a home and office Chandler in sync (the developers sometimes referred to this issue as "synchronization" and other times as "replication")? What if you wanted to mix "Chandler-native data" (information stored by Chandler itself) with data— say, email or addresses—from other programs?

Toy runs out of space on the whiteboard. "Let's just identify replication and foreign data as things that, right now, we don't have a design for, that we're doomed on."

Kapor rubs his eyes wearily. "*Mah nishtanah halailah hazeh*," he quotes the ritual Passover question. "How is this night different from all other nights, as they say? I've stepped into this movie at multiple points. It's *Groundhog Day*. How is this different?"

▮ ▮ ▮

One answer to Kapor's question sat across the room from him. Andi Vajda had interviewed at OSAF earlier that spring when the Chandler developers had begun to look for someone to write the program's email component. His interviews didn't wow them, but they kept his name on file. When they needed to hire a programmer to build the repository, Michael Toy thought he might be a good match.

Vajda is a bullet-headed, intense man with a twitchy blink and an accent marked by an Austrian family and a youth spent in France. Like just about every other programmer at OSAF, he had started playing with computers in high school. By the time he had graduated, he had hacked open the operating system for the school's minicomputer. "Basically," he later told me, "everything I've learned about operating systems, multitasking, memory usage, hard drives, file system layout, all of these things go back to

LEGO LAND ||| III

that. When I learned it again in college, it was old news. I thought, 'Yeah, I've seen this before.'"

When Toy outlined OSAF's repository quandary to Vajda, he had another "Yeah, I've seen this before" moment. He had spent nearly a decade working in the artificial intelligence industry for a company named Intelli-Corp, using the dynamic programming language favored in that field, Lisp. At IntelliCorp and later at two other companies he had been responsible for creating "dynamic object systems" that supported the flexible storage and retrieval of software objects.

"This is what I'm all about. If you don't give me an object system, I'll put one someplace, because I need one somewhere. It all goes back to Lisp. Coming to OSAF, I said, Ahh, it's another one of those. They need a dynamic object system. I had already done this several times. . . . It wasn't like, oh, this is going to be a neat research project, I'm smart, I can figure this out. I'd figured out a lot of it already."

There are decades of lore in the software industry about the "cowboy coder," the programmer who resists rules, prefers solitude, and likes to work on the edge. To a lot of managers, cowboy coders are a nightmare; to a lot of programmers, they are heroes. Though Andi Vajda didn't perfectly fit the cliché, there was something undeniably cowboyish about his approach to building Chandler's repository. His goal, he said, was simple: "Let's get these people unstuck."

■ ■ ■

For Vajda the May 21 meeting to slay the repository snake is serving as a sort of instant recap of the long debate at OSAF over how to store Chandler's data. He mostly listens, but every now and then he shows his hand.

"The heart of Chandler," Toy says, "is that data does not organize itself hierarchically."

"Oh," Vajda interrupts. "It will. When you have nothing, you need to start somewhere. You need a root."

"From the UI's perspective," Katie Parlante says, "there's no primary hierarchy."

"But there has to be something where you start," Vajda repeats. Maybe it's only for housekeeping purposes, maybe it's important to the machine but is hidden for the user, but Vajda is quite sure: You're not going to build a stable data store without some sort of hierarchy.

The next day Vajda takes the floor and begins diagramming his idea for a way to move past the repository standstill. The most important thing, he declares, is to get agreement on the Chandler "data model"—the definitions of the exact structures the program will use to store data. Once that's set, he can go off and start building a repository.

For the next few weeks a small group—Vajda, Anderson, Parlante, and Brian Skinner, a volunteer who had recently joined the project—met daily and agreed on certain things. The heart of Chandler's data model would, they reaffirmed, be the "item." An item could be virtually anything—to a user it might be a particular instance of an email, a note, or an appointment; but it could also represent part of the machinery of the program itself, like the definition of a "kind" of item ("email" or "note" or "appointment"). Items would, in time-honored object-oriented fashion, form a hierarchy of inheritance. Therefore, if each one of a group of items was defined as an instance of a certain "kind"—if they were all emails, for example—they would all inherit the characteristics of the "email kind" item. And all items would be stored in the repository.

This choice, to make Chandler "item-centric," had a couple of implications. One was that it moved the program decisively away from the world of RDF, where Chandler was born back when Morgen Sagen wrote its first repository. In RDF, the primary unit of data storage is an "attribute"; one Chandler item could contain lots of different attributes (for an email, typical attributes might be date, sender, subject line, body text, and so on). The difference between a more typical object-oriented item approach and RDF's attribute-based technique, the developers sometimes said, was like the difference between realms of physics: Items were like atoms; attributes were like subatomic particles. And as with the quantum world below the atomic level, the developers feared that an RDF-based data store would be unstable and unpredictable.

The other big consequence of the decision to make Chandler's repository item-centric was that it left Vajda free to take or leave ZODB. As long as he could guarantee that Anderson and the other Chandler programmers would be able to use the repository to maintain the relationships between their Python code and Chandler's data, it didn't really matter how he arranged for that to happen. And Vajda wasn't leaning toward using ZODB; he figured it would be faster, simpler, and better to use the same lower-level database that ZODB relied on, the Berkeley DB, and write his own code on top of it. This was pretty nearly what Lou Montulli and Aleks Totic had predicted OSAF would end up doing.

"I never even took a look at ZODB," Vajda later recalled. "There was intense time pressure. I could reuse ZODB, but it would take me a while to learn it. Or I could reuse my past experience in dynamic object systems and just do one. That was the faster route. Otherwise, I'd have to go and deconstruct ZODB, figure out how it does things, and then map Chandlerness onto it.

"And also there were a lot of steep requirements on the repository; it was going to have to do sharing and replication. I'm sure we could do them with ZODB, but if I'm in full control, I can do them quicker and maybe better, more appropriately for Chandler's needs. Another thing with ZODB was, it's Python. If tomorrow we need something that's not Python, we're hosed. I like to design stuff nimbly so that if the requirements change or something changes that we didn't expect, we can turn it around and say it can do that, too.

"It wasn't a political decision. It wasn't even a not-invented-here thing. It was sort of a survival decision. We said we need something really quick, now. I know how to do this. ZODB? I don't need it."

With the fundamental design of the Chandler data model in hand, Vajda was ready. In mid-June he left to spend a month in France with his family and relatives. By the time he returned, he had built the beginnings of Chandler's repository. Over the next couple of months he would finish the "first rev," or version. It was still missing a lot of its parts, but it had a usable top layer, or API—the key commands and data definitions the programmers needed to begin writing Chandler code that actually stored and retrieved items in the repository.

■ ■ ■

There's a brief scene in the movie *Jurassic Park*, sandwiched between tableaux of a rampaging Tyrannosaurus rex, that probably didn't make much of an impression on you unless you're a computer programmer. The little girl heroine is sitting in front of a computer screen in the control room of the dinosaurs-run-amok theme park, trying to figure out how to turn on a security system. She looks at it for a second and then exclaims, "This is a Unix system. I know this!" The day is saved, at least for the next ninety seconds.

Programmers in audiences all over the world cheered her—and not just because of the movie's tip of the hat to the operating system favored by university hackers and corporate data centers. They also cheered because they had all had similar eureka moments—instants of recognition when some urgent or intractable problem flashes into focus as something familiar, approachable, and manageable. *I know this problem. I've been here before. I can work with this.*

Andi Vajda showed up at OSAF, looked around, experienced such a moment, and began writing code. In a trice he dismissed months of wrangling over ZODB. A plug-in Lego part wasn't going to cut it. He was certain he could build it himself, better *and* faster. And he went off and made it happen. It was a little brash. It was also undeniably galvanizing. It pulled the Chandler team out of its software-time doldrums and back into the world of incremental progress.

It would be some time before the other developers at OSAF could know whether Vajda's confidence was warranted. But at least they had the beginnings of a repository. At least they were unstuck.

CHAPTER 5

MANAGING DOGS AND GEEKS

[APRIL—AUGUST 2003]

T he message was cryptic:

    1'WQE

One morning at OSAF in the summer of 2003, those characters appeared on my screen. I did not type them. Chandler did—Chandler the dog, that is, who had run into a meeting where I was taking notes and poked his snout onto my keyboard. I was startled, but I can't say I was surprised. Chandler was one of Mitch Kapor's two labradoodles, and the labradoodles had the run of the office.

Chandler and Cosmo, café-au-lait-colored and irrepressible, were the top dogs in a veritable office pound—including Siggy, a skittery white bichon frise, always looking for admirers; Petunia, a waddling, snuffling English bulldog who would sit guard by her owner's desk; Porter, a chocolate Labrador who was always on the prowl for chow; and a rotating canine cast of at least a dozen more. The Howard Street digs that OSAF had moved to from

Belmont in July 2003 were dog-friendly. That elated the many dog owners at OSAF and the related enterprises that shared the office space. Non–dog owners, who sometimes seemed like a minority, were not always quite as thrilled.

At that time labradoodles—a cross between a Labrador and a standard poodle—were right on the cusp of becoming a trendy breed. They made fine pets: "They get along well with other dogs," one description reported. "Extremely clever, sociable, and joyful. Quick to learn unusual or special tasks. Active, a little comical at times. Can attempt to outsmart their owners if undisciplined."

Was Chandler undisciplined? Well, he would sometimes leave his mark in yards-long wet trails across the carpet. One afternoon he chose a corner of the office for a deposit; not long after, the owner of the recruiting firm on the floor below wondered what was dripping onto her desk. Every now and then Chandler and Cosmo would lead a chase around the office's circular floor plan, into and out of the big meeting room where OSAF's developers might be hashing out some fine point of program architecture. If the room's entrance was blocked, the dogs would hurl themselves at the double doors like battering rams.

Mostly, it was fun to have the dogs around. They warmed up the place a bit, frolicking in blissful ignorance of the protocols and interfaces and data models that were on the minds of the nearby human beings. But as OSAF's staff grew and occupied a larger space, a laissez-faire approach to dog/programmer coexistence became problematic. Some sort of management structure was needed. A special "dog interest group" committee formed, under the aegis of OSAF's human resources manager, to deal with the matter.

There were inevitable trade-offs: Dog owners, and that included the boss, wanted to have their dogs around. Programmers needed to concentrate. But some of the programmers loved to work with their dogs around, while others found the distraction irksome. A dog-friendly workplace was a plus, but only if dog owners took responsibility for their dogs' escapades and accidents.

The dog interest group undertook to hold biweekly meetings and set up a special anonymous "dog channel" email address for reporting dog misbehavior. Some office spaces were officially designated as dog-free zones.

The trade-offs weren't easy. But how could you expect to accomplish anything without making some hard choices?

■ ■ ■

Having acquired his name from the nascent software project, Chandler-the-pooch would go on to lend his soulful face to the start-up screen of successive editions of Chandler-the-program in a sequence of William Wegman–style portraits: Chandler in a hard hat (warning—construction zone), Chandler in a lab coat ("now experimentally usable"), Chandler's face on a calendar (the program's calendar works!). The dog inevitably became mascot and symbol for the software project.

Quick to learn unusual or special tasks. Can attempt to outsmart their owners if undisciplined. Labradoodle or programmer? Say "managers" for "owners," and the overlap is striking. Typically, when people discuss the problem of managing programmers in animal terms, the comparison is not to dog training but rather to "herding cats." Programmers, the stereotype goes, are stubborn individualists. Repetition and rolled-up newspapers aren't going to do the job; they need coaxing and stroking, mice to chase, and bugs to swipe.

What dog management and software management share is the inevitability of trade-offs. When you choose to place one goal foremost—a disciplined pup? a deadline-meeting coder?—you find you must sacrifice something else. In the software world, most of the choices that arise boil down to one particularly heartbreaking three-way trade-off. Optimists call it the quality triangle; to pessimists it's the impossible triangle. Either way, it's usually bad news.

I first heard this principle from a veteran software developer and author named Dan Shafer who had come on board Salon in the early days to try to bring our ambitious but naive crew of journalists up to technical speed. When we scratched our heads about how slow work was going on Table Talk, an online conferencing system we had rashly opened before the software was working right, he would lean back in his chair, rub his buddha belly, and say, "There's an old saying: I can make it for you fast, cheap, or good. Pick any two."

You can't labor long in any engineering realm without encountering this painful formula, but it is in the world of software, with its infinitely plastic form, its difficult-to-measure costs, and its Slough-of-Despond-like sense of time, that the quality triangle has fully come into its own. And like all bad news in business, it invites pushback from managers determined to prove that their organizations can be the exception to the rule. They *can* have it all—good, fast, and cheap.

Programmers love grids. Here is how one might map out the quality triangle's choices:

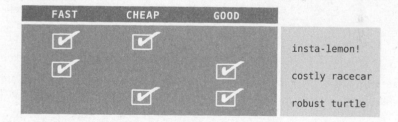

But this grid offers only an either-or picture, a set of absolute binary trade-offs. In practice, what managers face is more like a mixing board with linked sliders. You've decided to spend a little more money, easing the Cost lever over from "cheap" to "expensive"? That, you'll find, will nudge the Quality lever a bit farther from "crappy" and closer to "perfect." If you want to increase the speed as well, you'll have to push Cost even higher or push Quality back again.

But a manager gets to sit down at that console and move those sliders around only if a project is organized enough to respond predictably to decisions about cost and schedule and features or quality. Frequently, the work of managing software development is more primordial—an effort to summon order from a void, to answer questions like "Where do we begin?" and "How do we know when we're done?," and to marshal the energies of a refractory group toward a shared target.

■ ■ ■

One day some of the OSAF dogs got coated with some antiflea goop. Their owners attached little paper notes to the dogs' backs warning: "Treated for fleas—don't pet me!"

Kapor passed by, read one of the notes, and smiled. "Wouldn't it be great if people could do that, too?" he joked. "Wear signs like 'I'm grumpy today.' 'I've been bad.' It would take all the ambiguity out of human relations. I'm all for it!"

■ ■ ■

It wasn't long into the tangled arguments over Chandler's back end in the winter of 2002–3 that Kapor became convinced he needed to hire a manager. Schedules needed to be drawn up. Decisions needed to be made—and then made to stick. Processes needed to be adopted, adapted, or invented. He wrote on his blog, "We are already too large to work using the 'Vulcan mind-meld' methodology."

Kapor didn't want the job himself; he didn't think he would be especially good at it. As senior programmer and software architect, John Anderson had been partly filling this role, but he was diffident about assuming the nuts-and-bolts management tasks: drawing up schedules, trying to get programmers to meet those schedules, deciding what to do now and what to put off until later. He was more ruminative than gung-ho and might not be the best person for such responsibilities. And, anyway, he had his hands full making technical decisions and writing code.

Over the Christmas holidays at the end of 2002, Kapor posted to his blog under the heading "Making Design Decisions: Some Principles." The first item on the list read: "Implementation must be sequenced. It's not a real project until commitments are made to defer some capabilities. Doing everything at once is not an option."

"Everything" was a lot. The feature summary posted to the OSAF Web site as part of the project's original unveiling listed thirty-six separate features grouped into five categories. The email list alone contained seventeen features. Some—such as "fast searching via full-text indexing," "in-line viewing of attachments," and "user-defined views, rules, and filters"—were standard for any email program, although not necessarily simple to provide. Others— such as "user-scripting capabilities," "filing of messages in multiple folders," and "transparent encryption and authentication of mail"—represented complex innovations; they had either never been tried before or never been delivered satisfactorily.

Kapor and OSAF had said that Chandler would allow users to share information easily and flexibly in peer-to-peer mode, without relying on a server, and to synchronize multiple copies of the program and its data (for instance, between home and work). But Chandler also needed to offer security for users' vital information.

Could it do both?

Kapor and OSAF had promised that one day, Chandler would be a full-featured personal information manager, a PIM that handled email, calendar, addresses, to-do lists, and random notes. It would also be an "extensible developer platform," giving programmers an open-ended invitation to adapt Chandler to organize all sorts of things, including music, photos, blog postings, and anything else under the sun.

Could it be both?

Chandler's developers were going to operate in open source mode with their program built "iteratively," bit by bit, in public. But Kapor was also determined to produce a program that looked good, worked fast, and felt right.

Could they do both?

Nothing but trade-offs, everywhere you looked.

The project couldn't put all these choices off until the arrival of the new manager. That could take weeks or even months. "In the interim, we mustn't be waiting for a knight to ride in on a white horse," Kapor wrote in a memo to his team. And he was ready to resolve some of the trade-offs himself. As the repository debate dragged on, he had already made one big choice: He would front-load Chandler's basic PIM functions, like email and calendars, and put on hold the "developer platform" features, the open-ended aspects of Chandler that would allow programmers to adapt it for organizing all kinds of stuff. This was disappointing; after all, those were the very Agenda-like traits that had infused Chandler with excitement and charged up the online community. But they would have to wait. Instead of trying to do everything at once and inevitably failing, OSAF would try to focus on producing a good working PIM—with a handful of "killer features" that were "radically better" than the competition to "make adoption worthwhile," as Kapor put it in his Making Design Decisions posting.

But even narrowing the options to a few killer features still left OSAF with a wide-open field of choices to make, and maybe get lost in. Kapor posted this list of "killer feature candidates": "sharing; robust, transparent security; agent architecture; Agenda-like flexibility; [add your favorite here]." The final blank was a concession to reality—Chandler's planning still had a long way to go—and a statement of flexibility, of determination to seize the opportunity if a great new idea should come along. It was also an open door into which all manner of delays might walk.

■ ■ ■

The search for OSAF's software development manager did not go quickly. Although such a job carries authority, it is not one that many programmers aspire to fill. Some take their cue from the world of the Dilbert comic strip—with its barbs aimed at the archetypal, moronic "pointy-haired boss," or PHB, as acronym-loving programmers came to refer to the figure—and actively resist the whole concept, viewing anyone who takes the title of manager as an obstacle. (One programmer posting on a coders' site wearily put it, "The name for my pain is . . . managers.") Others accept the value of having good management but run screaming from the prospect of taking on the job themselves.

As the hunt dragged on, Lou Montulli and Aleks Totic suggested a name from their Netscape days. Michael Toy had been one of a band of employees at Silicon Graphics who left with its founder, Jim Clark, when Clark decided to start a new venture that would turn into Netscape. He had led the company's programming team through several hyperspeed cycles of the browser wars in an era that redefined the norms for software development, establishing new benchmarks for fast releases of new versions. He had what really was the only qualification that mattered when evaluating a software development manager: a track record.

Toy, a native of Livermore, California, had been programming all his life, from his student days at University of California–Santa Cruz and Berkeley, where he cowrote Rogue, an early Dungeons-and-Dragons–style computer game classic. He had built Rogue for the college campus's Unix-based minicomputers and was surprised to find, a few years later, how popular it had become. When first the IBM PC and then the Macintosh took the industry by storm in the early and mid-1980s, Toy and some friends started a company to port Rogue to the new machines.

"Once I had written my first computer program in high school, then all other things in the world seemed a lot less interesting," he recalls. "There was something really rewarding about the process of breaking problems down, explaining to the computer, watching the computer do what you told it to do. When I got out of college, it felt like—this is a common story I hear from programmers—the fools were *paying me money* to do what I would have done for free if only they had known." He pauses, deadpan. "That still continues today."

Though any early employee of Netscape ended up with a financial jackpot, those rewards weren't enough to hold its team together. Toy left as the company's suffocation by Microsoft (and its own mistakes) began to look inevitable, and took a break from the software business. In early 2003, he wasn't entirely sure he wanted to return, but the opportunity at OSAF fit every item on his dream job checklist: "I like the idea of working on software that millions of people will use, that will be on everyone's desktop. I think that the browser failed in its promise in a certain way, because a browser says that the only thing interesting to do with information is to look at it. . . . But it turns out that as soon as you have the browser, your next step is to organize and communicate information, and browsers never picked up that interesting problem even when it became obvious right away that that was going to be the deal. And so it's the problem that, since 1995 and a half, has been interesting to me to go solve. But I don't know about the pressure to do a crappy job in a hurry. It'd be nice if we had some luxury to do a better job, right? And it'd be great to do it open source."

Toy warned Kapor that he felt he wasn't very good at the personnel side of the job—coaching, performance reviews, and the like. Kapor told him not to worry; they'd work together on that stuff.

Toy started in March 2003. He was the polar opposite of the stereotypical Pointy-Haired Boss. He kept his hair long in back, pulled into a ponytail. He exuded a glum modesty. His personal home page on the Web featured a painting of himself as the Scarecrow from *The Wizard of Oz*, and in meetings he sometimes interrupted the people he supervised with requests like "Say that again for my small brain" or "Make Michael smarter." On his personal weblog he interrogated himself about his liberal Christian faith, and on the new blog he started at OSAF, he posted thoughts on the process of managing programmers. His OSAF blog, titled Blogotomy, carried this epigraph: "The transition from programmer to manager is often described as 'getting the lobotomy.' Here is a journal written by a now blissfully happy transformed individual."

On the Web site he maintained for a side venture called Undignified Labs—a funky recording studio for Christian musicians—he posted the following statement about himself: "Michael is the last person you would want to call to help you paint your house, but the first to call if you want to tear it down."

OSAF was doing neither of those; if anything, it was still trying to pour its house's foundations. One of the first things Toy needed to accomplish when he arrived was to nail down a development schedule. The O'Reilly conference in April set a deadline for the 0.1 (pronounced "dot-one") release of Chandler; after that, all the team had was a lengthy feature list and an open expanse of months leading to an eventual 1.0 ("one-dot-oh") release.

Kapor had decided to name the different versions of Chandler after locations in the Los Angeles area where Raymond Chandler set his mysteries. (Why not use the names of Chandler novels? Then they'd be stuck with a product labeled *The Big Sleep*.) So Chandler 1.0—an initial, full-featured version aimed at individual users—would be referred to as Canoga. A first version of the program intended for OSAF's new partners in the university world would be called Westwood. And a more distant Chandler 2.0, code-named Pasadena, would concentrate on building out the software so other companies could customize it for small- and medium-sized businesses.

Chandler 1.0, Canoga, had been promised by the end of 2003. But it was spring, and OSAF hadn't even settled on the basic architecture of its back end. For Kapor's programmers, as Toy would write on his blog, it felt like "the ground on which we are trying to stand is still cooling." How could they possibly meet such a deadline?

There was only one way to try, Toy knew: Make a schedule and start trying to follow it.

■ ■ ■

"Management," wrote Peter Drucker, the late business philosopher, "is about human beings. Its task is to make people capable of joint performance, to make their strengths effective and their weaknesses irrelevant." We're accustomed to thinking of management as the application of business school techniques that carry a scientific sheen: uniform measurements of productivity and metrics of return-on-investment. Drucker's definition sounds awfully squishy; he could be talking about an orchestra conductor or a stage director. But in emphasizing the art of management over the science, the human realm over the quantitative dimension, Drucker—who first invented the term *knowledge worker* and then offered invaluable insights into its implications—was trying to remind us that numbers are only a starting point for management, not its ultimate goal.

One great irony inherent in the management of software projects is that despite the digital precision of the materials programmers work with, the enterprise of writing software is uniquely resistant to measurement. Programming managers have struggled for decades to find a sensible way to gauge productivity in their field. The end product of a day's effort for the working programmer is code, and the most obvious yardstick of software productivity is the number of lines of code written. But it is an unsatisfying and sometimes even downright deceptive measure. Just as Noble and Biddle had found with their study of reusable software objects, there is nothing uniform about a line of code. There is no reliable relationship between the volume of code produced and the state of completion of a program, its quality, or its ultimate value to a user.

Andy Hertzfeld tells a relevant tale from the early days at Apple about his mentor Bill Atkinson, a legendary software innovator who created Quickdraw and Hypercard. Atkinson was responsible for the graphic interface of Apple's Lisa computer (a predecessor of the Macintosh). When the Lisa team's managers instituted a system under which engineers were expected to fill out a form at the end of each week reporting how many lines of code they had written, Atkinson bridled. "He thought that lines of code was a silly measure of software productivity," wrote Hertzfeld in his account of Macintosh history, *Revolution in the Valley*. "He thought his goal was to write as small and fast a program as possible, and that the lines of code metric only encouraged writing sloppy, bloated, broken code."

The week that he was asked to fill out the new management form for the first time, Atkinson had just completed rewriting a portion of the Quickdraw code, making it more efficient and faster. The new version was 2000 lines of code *shorter* than the old one. What to report? He wrote in the number −2000.

If counting lines of code is treacherous, other common metrics of software productivity are similarly unreliable. You can try to track program features or "function points," but they rarely divide neatly into units of similar difficulty or size; you end up making highly subjective calls about when a particular feature is done. You can try to chase bug numbers, measuring productivity by tracking progress in reducing the overall bug count. But sometimes finding bugs is essential work that brings you closer to finishing the job, even as it boosts the bug count. And bugs vary widely in the amount of time required to fix them, so your team might take a couple of weeks to whittle a list down from one hundred open bugs to twelve, only to find that the final dozen stubbornly resist a fix week after week. Furthermore, whichever unreliable yardstick you grab, you wind up facing the uncomfortable truth that Frederick Brooks reported in *The Mythical Man-Month*: Productivity varies wildly from one programmer to the next, frequently by as much as a factor of ten. So guessing how to staff a project can be as frustrating as estimating how long it will take.

Most software managers, well aware of these difficulties, end up improvising. There is a list of what needs to be done, subdivided into a series of tasks, and there is some method of keeping track of which of those tasks is (more or less) completed. Fully aware of the perils and paradoxes of software time, the manager will still expect individual programmers to try to estimate—or at least SWAG (take a Silly, Wild-Assed Guess)—how long each remaining task will take. The manager will then assemble all this information in one place, estimate a completion time for the project, and—if, as typically happens, the result suggests that the software will not be ready before the next millennium—start making trade-offs.

■ ▌ ▐

The informal approach to managing technical projects achieved its most famous incarnation at Hewlett-Packard with the concept of "management by wandering around," which was later popularized by Tom Peters's *In Search of Excellence*. This rigorous methodology requires managers to leave their offices, visit workers in their dens, and chat them up. "How's it going?" may not sound like the most incisive or analytical line of inquiry, but management by wandering (or walking) around was revolutionary in its way: It suggested that it was less important for managers to pore over ledgers than for them to get out and observe what their employees were actually doing. It placed a premium on human observation and contact.

But MBWA, as the tech industry's acronym-mongers soon dubbed the idea, doesn't translate well to the software realm: The work is simply not visible to the wandering managerial eye. No one has expressed this difficulty with more matter-of-fact precision than Watts Humphrey—a high priest of software management who led the IBM software team in the 1960s after Frederick Brooks's departure, and then went on to found the Software Engineering Institute at Carnegie-Mellon and to father a whole alphabet soup's worth of software development methodologies.

"With manufacturing, armies, and traditional hardware development," Humphrey argues, "the managers can walk through the shop, battlefield, or lab and see what everybody is doing. If someone is doing something wrong or otherwise being unproductive, the manager can tell by watching for a few minutes. However, with a team of software developers, you cannot tell what they are doing by merely watching. You must ask them or carefully examine what they have produced. It takes a pretty alert and knowledgeable manager to tell what software developers are doing. If you tell them to do something else or to adopt a new practice, you have no easy way to tell if they are actually working the way you told them to work."

All these blind spots mean that "the manager cannot tell where the project stands." Can't the manager just ask the team? "Most developers would be glad to tell their managers where they stood on the job," according to Humphrey. "The problem is that, with current software practices, the developers do not know where they stand any more than the managers do."

■ ■ ■

Humphrey's perspective is discouraging, but at least it's down-to-earth. How can you measure what you can't see? How can you judge a group's progress when its members don't know where they stand?

But in much of the business world today there lies, beneath the wide dissatisfaction with how software projects perform, a deeper anxiety—a fear that the root of the problem may lie not in failures of management technique but in the very nature of the people doing the work. To many executives, and even to their coworkers in sales or other parts of a company, programmers often seem to belong to an entirely different species. Communicating with them is frustratingly hard. They fail to respond to applications of the usual reward/punishment stimuli. They are geeks, and they are a problem.

Roughly a decade ago the word *geek* came into common usage to describe the kind of person who finds it easier to have relationships with computers than with other human beings. *Nerd*, the previously favored term, reeked too heavily of high school cliques and pocket protectors. The etymology of *geek* goes back centuries (a variant of the word even appears in Shakespeare's *Twelfth Night* and *Cymbeline*); the original meaning of "fool" or "simpleton" became localized in the early twentieth century to the circus sideshow, where you could pay to see the geek eat insects or bite off a chicken's head. As the personal computer moved to center stage of American business and culture in the 1990s, technical neophytes needing help connecting to the Internet or installing a new printer began seeking the help of the nearest "computer geek" they could find—without, at least at first, referring to them as such to their faces. *Geek*, like *queer* and other highly charged labels for outsider groups, began as a slur and only later was broadly embraced by its targets as a token of pride.

You can trace the evolution of the word *geek* through successive editions of the Jargon File, also known as the New Hacker's Dictionary, a compendium of programmer language and lore curated by Eric Raymond of "The Cathedral and the Bazaar." Early editions defined geek as "one who eats (computer) bugs for a living—an asocial, malodorous, pasty-faced monomaniac with all the personality of a cheese grater." More recently the definition begins on a note of higher self-esteem: "Geek: A person who has chosen concentration rather than conformity; one who pursues skill (especially technical skill) and imagination, not mainstream social acceptance."

As *geek* mutated its way toward respectability, it was joined by *geek out*. Author Neal Stephenson explained in a 2005 *New York Times* op-ed: "To geek out on something means to immerse yourself in its details to an extent that is distinctly abnormal—and to have a good time doing it." True geeks have a capacity to geek out on almost anything—even kitchen cleanup. An enthusiast of productivity software named Merlin Mann, who in 2004 started a blog called 43 Folders that quickly became a cult favorite among programmers, once composed a love letter to a book he was reading called *Home Comforts*: "Some deranged fold of my lizard brain gets most turned on by *nine detailed pages* on how to wash your dishes. . . . I think my wonderfully tidy (and endlessly patient) girlfriend has probably resigned herself to the fact that I only get really interested in something when it involves some kind of fancy system with charts and documentation."

While not all geeks are programmers and not all programmers geeks, the overlap between the two groups is thorough enough that their shared identity is assumed as a matter of course. Among programmers, geekery has a host of subclassifications: There are Linux geeks and Mac geeks, standards geeks and scripting geeks, database geeks and crypto geeks, each with their own subcultural reference points. The social spectrum of geekdom is fairly wide as well, stretching from the stereotypically buttoned-down engineer, who likes his pencils sharpened and his human interactions neatly defined, to the stereotypically unbuttoned hacker, with his gravity-defying hair and order-defying instincts. What they all share is a passion for specialized knowledge and a troubled relationship—at best clumsy, at worst hostile—with everyone who lacks that passion. You mean you don't know how to configure your home network? For every geek who will gladly do it for you and then show you six ways to tweak it should you so desire, there is another who will snap, "Just RTFM, dude!" at you and turn away without explaining himself, leaving you to discover via Google that you've just been told to *read the fucking manual.*

As long as programmers pursued their arcane labors in back-room obscurity, it didn't matter much that they were hard to talk to or get answers from. But as computers moved into every classroom and every home, and businesses of every stripe began to reshape their processes around custom-built software, the hands-off approach to geek management became less and less tolerable. Programmers found themselves with titles like Chief Information Officer, sitting in boardrooms or talking to the press. Communication with outsiders was a pain, but splendid isolation was no longer an option.

Why can't geeks and nongeeks just get along? One recent study by the Defense Acquisition University called "The Human Dynamics of Information Technology Teams" tried to get inside the heads of six hundred IT professionals (mostly programmers and their managers) in seventy-seven teams. Using a half dozen means of assessment, including observation, self-reporting, and a variety of surveys, the researchers found that the personality makeup of IT pros differs substantially from that of the general population:

> More than three quarters (77 percent) of our sample reported a preference for *thinking* decision-making, with only 23 percent preferring *feeling* decision-making. This is significantly higher than in the general population, where the split between these preferences is generally even. Thinkers, as they are termed, generally prefer logical, objective, impersonal decision-making, focused upon cause-effect relationships and the clarity that comes from objectivity (problem first, people second).
>
> Forty-one percent of the IT professionals surveyed reported being introverted thinkers (combination of

introversion and thinking preferences), nearly twice the percentage in the general population. Introverted thinkers often prefer a lone-gun approach to work, often avoiding teams, collaborative efforts, and the training that support such structures. This group is least likely to engage and connect interpersonally with others, and may avoid creating personal bridges of trust and openness with colleagues.

It would hardly surprise most programmers' coworkers to learn that their coder colleagues are more likely to be rational-minded loners. "A lot of people feel that communicating with the information technology professional is just slightly harder than communicating with the dead," Abby Mackness, an analyst with Booz Allen Hamilton who conducted the Human Dynamics study, joked as she presented its results to a crowd of defense contractor employees and consultants.

At the extreme end of the spectrum, the behavioral profile of the programmer—avoiding eye contact, difficulty reading body language, obsession with technical arcana—blurs into the symptom list of a malady known as Asperger's syndrome. Asperger's is defined as a mild form of autism, though unlike the more severely autistic, people who have it often have high IQs and can function on their own. Beginning in the 1990s, the rates of both Asperger's and autism, which had been rising in general, took an especially steep jump in Silicon Valley. A 2001 *Wired* magazine feature by Steve Silberman titled "The Geek Syndrome" examined why these conditions, which scientists believe have a high genetic component, might be concentrated so heavily among the offspring of programmers. As Silberman wrote:

Clumsy and easily overwhelmed in the physical world, autistic minds soar in the virtual realms of mathematics, symbols, and code. . . . The Valley is a self-selecting community where passionately bright people migrate from all over the world to make smart machines work smarter. The nuts-and-bolts practicality of hard labor among the bits appeals to the predilections of the high-functioning autistic mind. The hidden cost of building enclaves like this, however, may be lurking in the findings of nearly every major genetic study of autism in the last ten years. Over and over again, researchers have concluded that the DNA scripts for autism are probably passed down not only by relatives who are classically autistic, but by those who display only a few typically autistic behaviors. . . . The chilling possibility is that what's happening now is the first proof that the genes responsible for bestowing certain special gifts on slightly autistic adults— the very abilities that have made them dreamers and architects of our technological future—are capable of bringing a plague down on the best minds of the next generation.

Whatever the cause of the rising autism rates—whether Silicon Valley is conducting a vast experiment by inbreeding geeks or unrelated factors are responsible—there is something unsettling about the parallels between the autistic profile and the geek personality. Some observers of the programming tribe have suggested that in order to commune more closely with the machines they must instruct, many programmers have cut themselves off from aspects of their humanity. But the Asperger's/autism parallel suggests that, more likely, those programmers were themselves already programmed to hear machine frequencies as well as or better than human wavelengths. That can help them write effective code and design efficient algorithms. But it puts them at some disadvantage in understanding how to shape a program so it accomplishes its human user's goals. It also sets them up for trouble when it comes to the simple need to communicate with anyone else who isn't also a geek.

Around the same time that Frederick Brooks was writing *The Mythical Man-Month*, another programmer of that era, Gerald Weinberg, wrote a book titled *The Psychology of Computer Programming*. It was the first attempt to study the burgeoning new discipline with the tools of the anthropologist and psychologist. One of the things Weinberg found was that despite their reputation as loners, programmers *really needed the chance to talk to one another*—the more informal the setting, the better.

Weinberg recounts the saga of a university computing center where bureaucrats decided one day to remove the vending machines from a common area. The students who gathered there got too rowdy: Something Had to Be Done. In the days after the machines were carted off, the center's help desk personnel, who normally could handle their workload with ease, found themselves swamped. The lines stretched out the door. Yet nothing else had changed; there were the same number of students, courses, and assignments. The center's manager finally assigned a sociology grad student to interview the crowd waiting for help desk advice. The findings presented an early instance of what today's business scholars call the "watercooler effect":

> The typical behavior of a student when he arrived at the computing center was to pick up his output and head for the coffee machine. There, while sipping coffee, he could have a first look at the program and also show it to his buddies who might be standing around. Since most of the student problems were similar, the chances were very high that he could find someone who knew what was wrong with his program right there at the vending machines. Through this informal organization, the formal

consulting mechanism was shunted, and its load was reduced to a level it could reasonably handle.

The point, Weinberg commented, was that "informal mechanisms always exist and it is dangerous to change things without understanding them, lest you derange some smoothly operating system which you will not be able to replace at similar cost." In the decades since Weinberg's writing, managers have gone from clumsily wrecking such informal mechanisms to clumsily trying to encourage them. Office designers have begun purposefully including spaces intended to foster the hubbub that Weinberg's university administrators sought to squelch in their quest for order. But in the same time span, programmers have gone far beyond architects and managers: They have invented a profusion of technologies for staying in touch with one another, extending the software cosmos with multiple new genres of tools for coordinating a team's work.

These tools, sometimes collectively referred to as "groupware," include systems for instant messaging, chat rooms, bug trackers, source code version control, and the venerable email list. As OSAF began its work, two newer ideas, blogs and wikis, were enjoying their first flush of popularity. Each type of software can easily feed output into a different channel: Your source code control system, for example, can pop off an email message to a mailing list each time someone checks in a new code patch. Before you know it, you have dozens of channels to stay on top of, and staying in the loop can easily eat up much of a workday.

In the early days of programming, teams shared their knowledge on paper, in loose-leaf binders known as project workbooks, around which evolved complex processes of coordinating updates. (Brooks's *Mythical Man-Month* devotes many pages to these rituals.) But as soon as programmers could move off paper, they did. The original digital groupware was Doug Engelbart's NLS, the 1960s-era template for so much of what would unfold in the personal computing revolution to follow. Each new iteration of groupware has represented another stab at Engelbart's notion of bootstrapping—using the tools of computing to improve the effectiveness of developing computing tools.

At OSAF the mirror logic of this phenomenon was almost painfully vivid: As OSAF's programmers wrestled with different schemes for coordinating their work, it would become a rueful ritual for them to exclaim how much easier it would be to build Chandler if only they had Chandler itself.

Some observers have speculated that the avidity with which programmers pursue alternatives to in-person communication is a direct function of their aversion to sitting in a room and actually talking to one another face-to-face (FTF). It is undeniable that the typical geek dislikes management-mandated meetings that drag on and cut into actual coding time. But in fact the stereotype of the programmer as hermit applies only to a relatively small slice of the programmer population. The typical programmer loves to talk shop. Multiple species of groupware have proliferated not because programmers want to avoid talking with one another about their work FTF but because they can't shut up about it. More often than not, their colleagues today are in different time zones; in the typical open source project they may not ever share the same physical location ("meatspace"). In order to get anything done, they have had to find ways to get answers from one another at all hours, to spread those answers widely among their team, and to keep a record of those answers for the future.

■ ■ ■

Ward Cunningham was not looking to invent something big when he created the tool now known as the wiki; he was simply trying to make his own work a little easier. A veteran programmer and student of the arcane art of object-oriented programming, in the early 1990s he was an early and avid participant in the pattern language movement, an effort on the part of software developers to apply the ideas of architectural philosopher Christopher Alexander to their work. Alexander's book *A Pattern Language* derived a sort of grammar of construction by observing common elements or patterns in successful buildings. The software pattern–language people aimed to apply the same approach to programming. In the mid-1990s, they held a conference just as the Web burst into view, and Cunningham left it with an assignment: to build a hypertext repository that programmers could use to share their software patterns on the Web.

A decade later he told the story of how the first wiki came into existence:

> I figured, OK, I'm running this repository, send me your patterns in a text file and I'll run a little script, and I'll turn them into HTML [the language in which Web pages are coded]. I made simple rules, like a blank line means a new paragraph, and stuff like that. And I was amazed at how people who sent me files couldn't follow even those simple rules. I was three pattern documents into this thing, and getting pretty tired of it already. So I made a form for submitting the documents.

But people would submit a document and then pester Cunningham to make changes in the page after it was posted. So he added a button at the bottom of every page that simply read "Edit this page." You didn't need a password. Anyone could press it and start editing. "It was so simple, I was a little embarrassed to show it to people," he once said. "Could something so simple really work?"

Cunningham added one twist: If your page included words written with capital letters inside them—a style now known as CamelCase (ItMakesWordsLookLikeThis)—the software would look up those words in a database. If a page already existed by that name, the words would automatically be linked to the page; if not, it would create a *stub*, an empty template for a page, waiting for someone to fill it in. And since the wiki software stored every version of each page—tracking all the changes people made, much as programmers' source code control systems did—you could easily roll a page back to a previous state if someone made unwelcome changes.

Cunningham's pioneering wiki, the Portland Pattern Repository, grew over a decade to about thirty thousand pages. It inspired a whole wiki movement. (The fanciful name derives from a Hawaiian word for "quick" that Cunningham spied on a trip.) Its most celebrated offspring, the volunteer-built Wikipedia, has evolved, via open additions and emendations from anyone who chooses to contribute, into one of the Web's most widely known and used reference resources. Meanwhile, small companies sprang up to try to adapt the wiki approach to business. For years enterprises had labored to build intranets—internal Web sites for employees to share information—but their efforts tended to collapse under the weight of bureaucracy and complexity. Wikis seemed to offer a quick-and-dirty shortcut to the promised land of Web-based collaboration.

■ ▌ ▊

The Chandler crew had already begun to suffer from communication channel overload by the time Michael Toy arrived. When Kapor unveiled OSAF in October 2002, it opened up four public mailing lists on top of an existing private one. Kapor and several others had started their own blogs. Every day the developers shared a chat room via Internet Relay Chat (IRC); each day's IRC transcript was stored and searchable. The programmers used the CVS source code control system to store their code and check in changes to it without disrupting one another's work. They had Bugzilla to track bugs noticed and bugs fixed. And in January, OSAF opened a public wiki for sharing notes, research, and ideas.

From both a practical and a philosophical standpoint, Kapor strongly believed in transparency. OSAF's work and process ought to be as open to public scrutiny as feasible, he reasoned, both because Chandler's destiny as an open source project required it and because openness simply felt right. The programmers and managers hashed out most decisions, other than hiring and the most sensitive personnel matters, in full public view. That didn't turn out to be as difficult as traditional corporate practice might suggest. In the end, achieving openness proved less of a problem than achieving order and coherence.

Almost from the day the Chandler wiki opened its virtual doors, there were gripes about it. Some of the OSAF developers found the type-your-text-in-a-box interface clumsy and resisted using the wiki as a common workspace. Others took to it so avidly that their volume of daily notes quickly grew unwieldy. Everyone agreed that the wiki's just-add-a-page-any-where mechanism led to a mess: Newcomers didn't know where to begin, and regular users had trouble finding material they had previously visited.

The task of taming the wiki first fell to Kaitlin "Ducky" Sherwood, a programmer and email management expert who had joined OSAF to help shape Chandler's approach to email and organize the project's documentation (all the material that would help programmers and users understand the program's structure and features). Sherwood attacked the problem by dividing the wiki into a Chandler section for official documents, a Journal section for meeting notes and programmers' daily logs, and an anarchic, anything-goes "sandbox" area called The Jungle. Borrowing a verb from the realm of coding, the Chandler team referred to the process of wiki reorganization as "wiki refactoring." To programmers, *refactoring* means rewriting a chunk of code to make it briefer, clearer, and easier to read without changing what it actually does. Refactoring is often compared to gardening; it is never finished.

A year after Sherwood's big push to refactor the Chandler wiki in mid-2003, the developers had the same complaints: How could you easily tell which pages represented the most up-to-date information and which were obsolete? How could you get a simple overview of everything on the wiki? Where was its table of contents? If blogs and mailing lists relied too heavily on pure chronology for their structure, the wiki's anarchic format simply left too many users at sea.

■ ■ ■

Despite the multiple tools OSAF had deployed, Toy found they didn't provide him with answers to the most important questions he had as development manager: What had each programmer accomplished in the past week, and what task was next? Some of the developers were posting notes on wiki pages; others were dashing off the occasional message to a mailing list; others were posting to blogs. There was no one place to turn for an overview of the project's status.

Toy didn't want to solve the problem by adding yet another tool. At Netscape they had used Bugzilla to track non-bug-fixing work as well. Since it was designed specifically for tracking bugs, it wasn't ideal for general task management, but, he decided, it was better than adding yet another system to the ones OSAF's team already had to mind.

Some of the programmers were perfectly happy with Bugzilla. Others found it irritating and chafed at the time it took to enter their tasks via its Web-based forms. John Anderson, in particular, resisted the approach; he could spend all day entering his tasks into Bugzilla, he complained, and never get any coding done. And for the nonprogrammers at OSAF who also needed to provide regular updates on their work, Bugzilla was especially unfriendly.

For six months Bugzilla remained OSAF's official project management tool, but as willingness to use it grew increasingly sporadic, Morgen Sagen began working on a homegrown Status Manager for the team—a Web-based tool that would streamline the process for entering tasks and viewing them sorted by person, by project, by time, and by status. When Sagen ultimately unveiled the new tool in November 2003, it was a big hit among the developers. If they couldn't use Chandler itself to organize their work, the Status Manager was a good stopgap.

Still, the time Sagen had spent building the Status Manager—like the weeks Ducky Sherwood devoted to weeding the wiki—was time not devoted to building Chandler.

■ ■ ■

In *The Mythical Man-Month*, Brooks identified the Tower of Babel as "man's second major engineering undertaking" (after the Ark) and "first engineering fiasco." Babel's managers, Brooks declared, had all the "prerequisites for success" a modern-day management consultant could wish for, including a clear mission, manpower, materials, plenty of time, and adequate technology. Babel failed because its builders "were unable to talk with each other; hence they could not coordinate. When coordination failed, work ground to a halt."

In software management, coordination is not an afterthought or an ancillary matter; it is the heart of the work, and deciding what tools and methods to use can make or break a project. But getting sidetracked in managing those tools is a potent temptation. When the cry of "Let's build it ourselves!" arises, geeks are all too happy to rally and cheer. A celebrated (and perhaps apocryphal) bit of graffiti from MIT captures this: "I would rather write programs to help me write programs than write programs." Similarly, there is a saying attributed to Abraham Lincoln: "Give me six hours to chop down a tree, and I will spend the first four sharpening the axe." This principle, which found its way into the business advice manual *The 7 Habits of Highly Effective People*, appeals to every programmer's passion for toolmaking. But if it becomes an end in itself, it can drive the best-organized project into a ditch. James Gosling, the programming guru who invented Java, discussed this predilection in his blog:

> There are a couple of dark sides: The one that people talk about most is that it's important to actually finish sharpening the axe and get back to chopping. Often tool building is far more fun than actually doing the job at hand.
>
> But for me, the big problem with "axe sharpening" is that it's recursive, in a Xeno's paradox kinda way: You spend the first two thirds of the time allotted to accomplishing a task actually working on the tool. But working on the tool is itself a task that involves tools: To sharpen the axe, you need a sharpening stone. So you spend two-thirds of the sharpening time coming up with a good sharpening stone. But before you can do that you need to spend time finding the right stone. And before you can do that you need to go to the north coast of Baffin Island where you've heard the best stones for sharpening come from. But to get there, you need to build a dog sled. . . .
>
> Xeno's paradox is resolved because while it is an infinite sum, it's an infinite sum of exponentially decaying values. No such luck in this case. The tool building jobs just get bigger, and you pretty quickly can lose sight of what it was you started to do.

Another term programmers have applied to this phenomenon is *yak-shaving*, defined in Eric Raymond's Jargon File as a "seemingly pointless activity which is actually necessary to solve a problem which solves a problem which, several levels of recursion later, solves the real problem you're working on."

Programmers will always be able to justify any individual act of axe-sharpening or yak-shaving as "actually necessary." Some of them certainly are. It is the hard lot of the software manager to decide at what point the axe-sharpening or yak-shaving has lost touch with a project's original goal and to summon the programmers back from their happy tool-tending sidetracks to the primary task.

■ ■ ■

OSAF had released Chandler 0.1 in April and spent much of May in a series of snake-slaying meetings. The urgent decision Michael Toy now faced was how to schedule the 0.2 release. Should they pick a target set of features and declare the release done when those features were all coded and working? Or should they set a firm date and declare that the portions of the program that worked as of that date, whatever they turned out to be, were Chandler 0.2? Toy's instinctual preference was the latter choice—"clock-driven release" or "milestone by date." After all, this was still an alpha product; no one was using it or could use it. And if you didn't pick a line-in-the-sand date and stick to it, you risked the sort of slippage that had carried Chandler 0.1 into extra innings.

Toy and Kapor finally agreed that OSAF would follow a clock-driven schedule of quarterly releases. It was already June; Chandler 0.2 would be due in September. With Andi Vajda on board, the repository was finally beginning to come together. The programmers had their tools lined up, their axes sharpened.

All that was missing from Chandler was a finished design.

CHAPTER 6

GETTING DESIGN DONE

[JULY–NOVEMBER 2003]

L ate one night in my research for this book, as I plunged deeper into an awareness of all the different ways a software system can go awry, I said to myself, "Don't be an idiot. Back up your work."

So I devised a backup plan. The first step was to gather the fruits of my labor into one big folder: Book Project. Into Book Project went all my saved files from my office computer and home, all my Web addresses and interview notes and midnight jottings.

There! But now Book Project itself was buried three levels down in a folder hierarchy. That was inconvenient. I was going to be spending a good part of my work life in this folder for some time, so I wanted to move it to a handier spot. I opened up an Explorer window, the basic tool for file navigation in Microsoft Windows, and found Book Project. I clicked on it with my mouse, held down the mouse button, and began to drag the file folder to—

Well, where, exactly? I hadn't thought this through. It was important. There were too many options. I needed time!

I dragged the mouse pointer off the screen and released the button— assuming, based on some sense-memory of how other programs worked, that this action would make the folder snap like a rubber band back to where it started.

It did not. The Windows programmers who'd had to anticipate such an event assumed something different. If you dragged a folder off the screen, they figured you wanted it to disappear.

Book Project was gone. All of it. Deleted.

I stared numbly at the screen. How did *that* happen?

Okay, no need to panic. Check the Trash; it must be in there. Just drag it out again.

Not there.

Book Project was no more. It had ceased to be. It was an ex-folder.

My pulse raced. My temples throbbed. One detached sliver of my mind pondered the irony: All that research into software development wrecked by a software mistake! But the rest of me was an adrenaline-infused mess.

I started furiously researching undelete utilities on the Web. I'd never needed one before. Then I froze. If you need to recover deleted files, I remembered, you don't want to use your computer in any way that might cause it to write new stuff to your hard disk. The new files might write over the remnants of the ones you were trying to recover.

Finally, I calmed myself, and a little voice in my head piped up: *Undo.*

Right. Of course. Undo. I looked under the Edit menu in Explorer. There it was—my savior. One click on Undo, and my Book Project baby was back, safe in its nest in the folder tree.

I paused a minute to give silent thanks to the Windows programmers who had made this recovery possible—even though they were probably the same ones who had set me up for disaster. Then I went to sleep.

Creating end-user application software—software intended for use by mere mortals—means anticipating myriad combinations of human actions and machine reactions. A computer screen has hundreds of thousands of pixels; a user can click on any of them at any time, and a program must know what to do. It turns out there are more things under the sun that crazy users do than even the best programmers can possibly imagine.

All sorts of conventions have evolved to help tame and organize this sea of possibilities, like the difference between single clicking and double clicking, or the idea of "focus"—the part of the screen that is active in the foreground. (In the typical email program, for instance, your focus might be on the list of new messages at the center of the screen, but if you click on the side panel that displays a list of your mail folders, you switch focus to that screen element, and that is where your mouse clicks and keystrokes will be fed.) But these are conventions, not laws of nature; there is nothing innate about them. They evolve as generations of software come and go. And just as in biological evolution, fragments of an older system sometimes survive as artifacts in a new one.

Anyone who learned to use computers in the days of the command-line interface understands the simple notion that most of the time the computer isn't going to "read" the input you type until you hit the "enter" key at the end of a line. It is a reflex; we don't even think about it. So when I introduced my parents to the Web nearly a decade ago, I showed them how they could type any Web address into the bar at the top of the browser. When they tried, though, nothing happened. I had forgotten to tell them, "Hit 'enter' when you're done." The Web browser is a mouse-and-windows, GUI-style program, but that address bar is a little vestige of the command-line era, carrying an old set of customs and behaviors with it.

By default, it is usually programmers who end up trying to figure out how a program should react to each of the thousands of possible combinations of user inputs and machine states. But they are not always well equipped to put themselves in a user's shoes and imagine what will make sense. They spend the bulk of their working hours wrestling with digital minutiae, and their reflexes have already been trained in the customs of the systems they build. Concepts they take for granted are often entirely alien to nonprogrammers; users' assumptions may well be foreign to them. (One work-around that programmers often rely on is the "mom test," in which the presumed-to-be-clueless parent serves as a hypothetical use case or sometimes an actual guinea pig.)

Beyond that, it is the programmer's job to imagine extreme possibilities and unlikely scenarios. What if the power goes out while you're in the middle of saving your work? What happens to a calendar when the user has set his computer's clock to the year A.D. 2099? (Microsoft Windows allows you to do this, and when millions of people use a piece of software like Windows, sooner or later someone will end up making every possible un-likely choice that the software allows.) Programmers call these *edge cases*, and they are often where bugs hide. Trained to imagine them with exhaus-tive thoroughness, programmers' pragmatic, gut-sense awareness of which cases really matter can get rusty. They spend so much brainpower out on the edge that they lose sight of the center.

If a programmer isn't the best person to make the decisions that collectively constitute a software program's design, then who is?

Mitch Kapor has always been passionate about the nuts and bolts of software design, from the broadest philosophical questions to the most mundane details of fit and finish (getting an application to look and feel right to a user). Establishing software design as a discrete profession has been one of his causes for nearly two decades.

In 1990, at the PC Forum gathering of computer industry luminaries, Kapor first delivered the text of his "Software Design Manifesto."

> No one is speaking for the poor user. There is a conspiracy of silence on this issue. . . . Scratch the surface and you'll find that people are embarrassed to say they find these devices hard to use. They think the fault is their own. . . . Everyone I know (including me) feels the urge to throw that infuriating machine through the window at least once a week. . . . The lack of usability of software and the poor design of programs are the secret shame of the industry. . . . What is to be done? Computing professionals themselves should take responsibility for creating a positive user experience. Perhaps the most important conceptual move to be taken is to recognize the critical role of design, as a counterpart to programming, in the creation of computer artifacts. And the most important social evolution within the computing professions would be to create a role for the software designer as a champion of the user experience.

Software design, Kapor argued, was not simply a matter of hanging attractive graphics atop the programmers' code. It was the fundamental creative act of imagining the user's needs and devising structures in software to fulfill those needs. As the architect draws up blueprints for the construction team, so the software designer should create the floor plans and elevation views from which software engineers would work. Reaching back to ancient Rome, Kapor proposed applying to software the architecture theorist Vitruvius's principles of good design: *firmness*—sound structure, no bugs; *commodity*—"A program should be suitable for the purposes for which it was intended"; *delight*—"The experience of using the program should be a pleasurable one."

A lot of Kapor's manifesto may sound self-evident today, but his assertion that designers should take the lead in creating software remains controversial. Some programmers look down on the very idea of design and designers, and are quite certain they are the best arbiters of how their programs should look and behave. Others believe in the primacy of code, feeling that you can't really understand anything about the shape or workings of a piece of software in the abstract. Therefore, "Design first" (also known as Big Design Up Front, or BDUF) can't work. So it's better to get a program up and running fast and then refine it in response to user feedback.

Kapor never maintained that design had to happen separate from and before coding. "During the course of implementing a design, new information will arise, which many times will change the original design," his manifesto declared. "If design and implementation are in watertight compartments, it can be a recipe for disaster because the natural process of refinement and change is prevented."

Kapor himself had always been the lead designer at OSAF, but everyone there had some form of the design religion. Hertzfeld's dedication to usability verged on the obsessive. It was a rare exchange with him that didn't lead to an observation like "The user isn't going to get that" or "Let the user decide."

But by the summer of 2003, OSAF still did not have a fully worked out design for Chandler; the program remained more a set of principles and aspirations and commitments than a batch of blueprints. Part of the problem was that Kapor and his team had had trouble hiring a designer. They needed someone with a good graphic sense who also understood interface design and could work closely with programmers. Kapor's standards were exacting. They simply hadn't found the right person yet.

But it was also hard to nail down Chandler's design because doing so meant resolving difficult trade-offs between the program's ambitions and technical realities. Making firm technical choices was hard in the absence of a settled design, and settling on a design was hard in the absence of a technical roadmap. The work kept bumping into this chicken-and-egg problem, and it cast a pall of discouragement over much of the OSAF team.

Disheartened by the slow pace, Lou Montulli and Aleks Totic, the Netscape kids, drifted away from the project they had volunteered for six months before. Montulli later explained his decision to leave as a case of a bad personality fit: "I'm a fairly disruptive person. I like to push change fairly rapidly. Throughout the organization, people wanted to take a much more cautious, slow approach to software development, which didn't feel right to me." Totic turned his energies toward a new project, adapting an increasingly popular open source development toolkit called Eclipse to work with Python. Montulli returned to Lake Tahoe to plot new start-up ventures.

■ ■ ■

In July 2003, as Andi Vajda began digging the foundations for the Chandler repository, Mitch Kapor established an informal working group with John Anderson, Andy Hertzfeld, and later Michael Toy to hash out a plan for building Chandler's user interface. Work on the pieces—or parcels— of Chandler, like Katie Parlante's calendar, had reached a standstill. The program was missing a key part, a structure that was at various times called "the big table," "the summary table," or "the superwidget"—the piece of the program that would present your email inbox or your list of tasks or calendar events in a brief digest or summary form.

Someone had to step forward and begin coding the superwidget. Hertzfeld had written a version of such an interface for the Vista prototype, but he felt that for the real Chandler the job belonged to John Anderson. Anderson agreed. But he didn't want to plunge too deeply into the code until he was confident that he understood how Kapor wanted the superwidget to work and until Chandler's key contributors had reached a consensus on the proper approach to building it. And he had some ideas of his own about how it should be done.

■ ■ ■

"The question is," said Mitch Kapor, deep in the middle of a long meeting in a series of long meetings, "How do we sequence things to avoid spending an infinite amount of time before anything useful happens?"

"It's only infinite if you're stuck in a loop," Hertzfeld replied.

■ ■ ■

As the conversations started, Kapor, Anderson, and Hertzfeld decided that both users and programmers would work with Chandler primarily through what they began to call "Chandler documents." Like a word processor document, a Chandler document would store data, but like a spreadsheet document, it would also be a kind of template, storing code and rules for handling that data. And as with spreadsheets, the field of software innovation that formed the cornerstone of Kapor's career, Chandler documents would put the kind of power normally reserved to skilled programmers—the ability to extend the functions and uses of the program—directly into the hands of everyday users.

With the new focus on documents, the group's discussions began to be known as Document Architecture meetings. They happened throughout the summer, sometimes for three or four all-afternoon meetings a week. Though their goal was to resolve open issues so that Anderson could start writing a superwidget, they found it difficult to narrow the focus of their talks. A consideration of the ways people might hypothetically accomplish tasks with Chandler would take a left turn into a discussion of agents, Andy Hertzfeld's long-held vision of automated, semi-autonomous software routines that would handle routine tasks for the user. A debate over how much fine control users should have over the look and layout of Chandler on screen would quickly turn into a free-for-all about Chandler on mobile devices or Chandler access over the Web. And every discussion seemed to wind up back at the idea, theoretically discarded but still gnawing, that Chandler might somehow benefit by adopting or adapting parts of Mozilla.

It had been nearly a year since Anderson first examined and rejected Mozilla's tools, which had evolved out of Netscape's software when Netscape opened up its code in 1998. Since then, the influx of former Netscape employees—from Totic and Montulli to Toy—had made "Should we look at Mozilla again?" into a regular OSAF refrain. But what really brought the question back into the spotlight was Mitch Kapor's new role as chairman of the Mozilla Foundation.

Originally, Mozilla's core programmers mostly held jobs on Netscape's payroll, and when Netscape got bought by America Online in 1998, they became AOL employees. By 2003 that meant they were employees of Time Warner, the media conglomerate that had merged with AOL at the height of the dot-com boom and later came to regret the move. Deep in the doldrums of the tech downturn, AOL executives began asking: Tell me again why we're paying a bunch of programmers to give away code? The future of Mozilla looked increasingly uncertain.

Kapor already had one connection with Mozilla through Mitchell Baker, who had been working at OSAF part-time since late 2002. Another Netscape veteran, Baker had played a central role in the open sourcing of the Netscape browser: She wrote the license under which the Netscape code was released. AOL, as its support for Mozilla grew cold, had laid her off in 2001, but she continued to help lead the Mozilla Project as a volunteer. Kapor asked her to join OSAF and use her open source experience to help his organization build a community of developers around Chandler. She also kept Kapor apprised of Mozilla's increasingly precarious position inside AOL.

In May 2003, a chance meeting between Kapor and AOL executive Ted Leonsis at an industry conference led to a solution that left everyone happy: AOL would divest itself of the entire Mozilla operation, but not without helping fund a new nonprofit foundation (with $2 million over two years) that would take over the payroll for key Mozilla personnel and carry on the project's work. Kapor kicked in an additional $300,000 and agreed to serve as the foundation's chairman.

Since the Chandler developers still hadn't gained much traction and Kapor now had a formal role with Mozilla, Kapor suggested it made sense to look at Mozilla again. There was, he felt, at least an opportunity for a "grand synthesis."

One afternoon in late July 2003, the document architecture group gathers in the small conference room at OSAF's Howard Street office. Anderson opens his preferred personal information manager—a graph paper notebook. Hertzfeld has his, a ledger-style book with a red ribbon bookmark. Next to Kapor lies a yellow legal pad. Toy is the only one with a laptop computer. Everyone has grabbed a slice of pizza from the kitchen area of the new office.

Hertzfeld, impatient to move Chandler along, proposes a radical idea: Mozilla, he points out, is already structured to incorporate other programs as plug-ins. Why not build Chandler itself as one big Mozilla plug-in? Of course, he admits, there'd be problems. A browser-based design would certainly require a lot of rethinking of Chandler's goals. But relying on the browser's interface would save the programmers enormous amounts of the labor involved in building a new interface themselves.

"There's so much work ahead of us. It would be great to strap on some booster rockets," he says.

Michael Toy had often brought up the Mozilla option himself, but this time he raises cautions. "It's been forty thousand years since the invention of the Internet, and we still don't have a way for dumb people to make Web sites that are useful. And a Web browser is not a very good interface to something that is not the Web. It just seems like we'd be strapping a bad backpack on before we start walking."

Comparing Mozilla with the wxWidgets tools that OSAF has already begun using, Anderson says, has been a frustrating apples-and-oranges exercise. "I feel like I'm randomly roaming around the top of an iceberg."

Kapor gets up, goes to the whiteboard, sketches a simple layout of rectangles, and begins to label them. "Here's the screen. Certain areas sitting inside these rectangles are definitely going to have these rich widgets where there's sophisticated user interaction and it's tied to the data model. Tables. Outlines. Different views. That's our special flavor. We know we're going to have those things. These things have to be laid out. And you've got other standard GUI stuff—ornamental text and buttons and controls. So the question is, where does *this* come from?" He points to a spot where he's written *layout—base widget set.* "Our plan of record is this comes from wxWidgets. And there's a question: How much work will that be?"

Anderson speaks up: "Right now, sizers would be the simplest way to do layout." (Sizer code is the basic means the wxWidgets toolkit uses to lay out window shapes onscreen.)

"This is proven to be an order of magnitude harder for people to do than putting up Web pages," Hertzfeld objects.

"One of my concerns about sizers is that they're two-dimensional," Toy says. "CSS [Cascading Style Sheets, the layout specification format that modern Web browsers use] lets you put things on top of things."

As they talk, Kapor is writing on the board: *Is sizer framework sufficiently flexible? Expressive?*

"Part of the idea of using Mozilla was that it might give you for free some of the expressiveness in layout," Kapor says.

"A lot of people find CSS pretty daunting," Anderson says.

"You'd need user-friendly editors," Kapor says. "Is the absence of those contingent or fundamental?"

"Michael's opinion," Anderson replies, "is that it's fundamentally hard to make a good GUI for CSS."

"It might be that we just arbitrarily decide not to solve the rich document problem," Toy says. "I can feel that as a black hole."

Kapor writes again: *What part of the rich document problem to solve?*

At one point or another, each of the participants in the meeting had admitted feeling a sense that they were "handwaving"—talking in generalities without rigorously working out details. The biggest wave of the hands seemed to surround the meaning of the term *document* itself.

"Could you do a whole app as a document?" Kapor asks.

"It's could you do it as a *set* of documents," Hertzfeld says.

"Is Chandler a document?" Toy asks.

Hertzfeld replies, "We talked about, are dialogues documents? Are they part of the base document projected into a window? Or are they a separate document?"

"Then," adds Anderson, "there's also the chrome" (the frame surrounding the application windows where users open menus, manipulate scrollbars, and so forth).

"Is there free will?" Kapor jokes. "There are lots of questions, but given that they have a metaphysical component, I prefer to defer them till we're at death's door."

The four men begin to talk about how a Mozilla-based Chandler might provide a simple feature like a "progress bar," an empty bar that gradually fills up, giving the user a real-time indication of progress made on some task like receiving a big pile of email messages. This sort of dynamic interface element, which updates in real time, is the kind of feature we take for granted in most everyday programs. It might be a stretch for a Web-oriented platform like Mozilla, which was designed to display data in static screen-fuls delivered from a server. A Mozilla-based Chandler would need to base such a feature on Javascript, a programming language meant specifically for embedding program code inside Web pages. (HTML, the basic language for publishing Web pages, isn't really a programming language at all; it just tells a browser what to show on screen.)

"You never see this in a Web application because the client can only make a few round trips to the server in a second," Kapor says. "I want a sense of confidence that we'll have a first-class way to do this. To me this is at the heart of the whole PC-versus-Web thing. You can have only so many round trips to the server per second or your performance gets clobbered."

The pizza is cold. The discussion has consumed the afternoon. Kapor says he wants to set up a meeting for the group with Brendan Eich, Mozilla's lead architect (at Netscape he'd created Javascript).

"If there is a huge advantage in us integrating with Mozilla, it's at a level too subtle for us to identify yet. But I want to see whether it's a worthwhile conversation," he concludes. "Is it major surgery? A double lung and heart transplant is radical and possible. If it's a head transplant, it's not worth talking about."

Whatever you do, Kapor tells his colleagues, don't let this become a blocking issue. And settle it within a month.

■ ▌ ▐

A month later the issue was indeed resolved. The document architecture group abandoned the idea of building Chandler on top of Mozilla and decided to stick to its original plan. The term *document* remained elusive, and as the focus of the work narrowed to the nuts and bolts of Chandler's layout engine—not "What will the layout of Chandler be?" but "What tools will lay out Chandler, and how will the programmers manipulate them?"—they came up with a new label for their work. The system they were laboring over would be called CPIA (they pronounced it like the color *sepia*) for "Chandler Presentation and Interaction Architecture."

CPIA would be built on top of wxWidgets, but its goal was to wrap Chandler's Python code into packages called "blocks" that would be connected to each wxWidgets element. In the distant future, the idea was, CPIA blocks would make it easy for nonprogrammers to build new additions to Chandler by simply hooking preexisting blocks together. On a closer horizon, CPIA blocks would make it easier for OSAF's programmers to build Chandler in the first place. The program would store its blocks as data in the Chandler repository itself. This "data-driven" design would theoretically make it easier to change the behavior of a block; instead of writing new program code, you could just make and store a change in the data.

CPIA was a specific instance of the Lego Land dream of reusable software parts. The CPIA design document, which evolved on the wiki over many months, declared, "Blocks are designed to fit together like Lego blocks." For John Anderson the ideas behind CPIA constituted a long-held vision of a better model for programming. With Chandler he had a chance to make that dream real. Kapor was in sync with this dream; in an early interview with me, he had described his vision of building Chandler out of snap-together Lego parts, "top to bottom, inside out."

For programmers, the Lego concept had an irresistible gravitational tug. But the OSAF team didn't grab someone else's Lego set and begin snapping blocks together; they would end up fabricating their own set from scratch, one custom-made block at a time.

Peter Drucker, the business philosopher, offers a parable about work:

A favorite story at management meetings is that of the three stonecutters who were asked what they were doing. The first replied, "I am making a living." The second kept on hammering while he said, "I am doing the best job of stonecutting in the entire country." The third one looked up with a visionary gleam in his eyes and said, "I am building a cathedral."

The third man is, of course, the true "manager." The first man knows what he wants to get out of the work and manages to do so. He is likely to give a "fair day's work for a fair day's pay."

It is the second man who is a problem. Workmanship is essential, . . . but there is always a danger that the true workman, the true professional, will believe that he is accomplishing something when in effect he is just polishing stones or collecting footnotes.

Kapor called the CPIA work an "infrastructure investment." But it looked suspiciously like axe-sharpening or yak-shaving. Were Anderson and his colleagues "just polishing stones"? There was no way to know in advance.

◼ ◼ ◼

Andy Hertzfeld and John Anderson are sitting in a tiny conference room at the Howard Street office one August afternoon. They talk about Hertzfeld's ideas for agents. They talk about "contexts," a new notion of organizing Chandler data around related groups of user activities. They talk about Anderson's progress in assembling a flexible interface-building kit. Then they look at each other.

"I'll be excited when we're moving forward on implementing," Hertzfeld says. "My own style is, I implement too soon. But maybe that will cause the group average here to come out right."

"I try to play all the chess moves in my head before I start," Anderson answers.

Hertzfeld laughs. "Bill Atkinson used to say about me, everything was 'Ready, fire, aim!' "

"I've typically done most of a project by myself," Anderson says. "I haven't been as good at splitting it up so that lots of people could work on it together."

Even, Hertzfeld asks, at Next?

"When I was a manager there, I wasn't writing code. It's very different when you're one of five people doing one fifth of the work. I've either done the bulk of it or been managing it."

Anderson pauses. The ventilation system whirs. "And I have a little bit of a perfectionist thing." He closes his notebook and gets up. "What I really want to do now is go work on this!"

■ ▌ ▌

In the world of software, *integration* means taking a body of code that works fine by itself and connecting it to the other existing parts of a program that have in turn been working fine. The integration point is typically where a software project hits big trouble. Chunks of code that worked well enough separately often balk when asked to work together; they fail to hook together correctly, send messages that can't be interpreted, or stubbornly refuse to start or stop. (This is one of the many ways in which programmers' products sometimes mimic their human creators.)

Chandler was now rolling toward one such integration point. While the

CPIA gears cranked up in the background, Michael Toy began pushing Chandler's developers down the tracks he had laid out earlier that summer toward a "clock-driven release": Chandler 0.2, scheduled for late September. OSAF was already in the practice of releasing new milestones of Chandler every week or two, but a "dot release" was meant to be a larger event. So as the date for 0.2 drew closer, the developers began to try to take the work they had been pursuing on their own—like Andi Vajda's new repository and Andy Hertzfeld's work on agents—and add it to the main trunk of Chandler code.

Most projects today embrace the idea of continuous integration: The programmers always keep their latest code checked in to the main trunk of the code tree, and everyone is responsible for making sure that their new additions haven't thrown a spanner in the works. Later on, OSAF would end up achieving a higher level of continuous integration, but for 0.2 the process was more like what software-development analysts call "big-bang integration": all the programmers try to integrate their code at the end, and everything breaks.

Toy knew the dangers, but he felt they had made a collective commitment to release a version of Chandler every three months and ought to meet that deadline. Though Vajda's work on the new repository was nearly ready for integration, there wasn't quite enough time in the schedule to include it all in 0.2. Only some of the supporting technical framework for Hertzfeld's agents was ready. CPIA was barely in the blueprint stage.

But a deadline was a deadline, so the development team did what it had to in order to prepare a new public release of Chandler. After the "code freeze" date, they stopped checking in any code related to new features, focusing instead on fixing outstanding bugs and creating some documentation (explanatory material for other developers and users) for the release.

Chandler 0.2 was unveiled on September 25, 2003, close to a year after OSAF had first announced Chandler to the world. It arrived under a cloud. The few users who downloaded it and tried it out were surprised to see that it actually did even less than Chandler 0.1 had. It was like the shell of a structure that had been gutted and only partially rebuilt.

The developers, particularly Vajda, felt frustrated. "Chandler 0.2 is about to ship, and I beg you not to download it," Michael Toy wrote in his blog on September 19. "I'm serious. Don't download it, don't tell your friends about it, don't add links to the download page to a Slashdot posting." Toy mused about the wisdom of even trying for a formal release of such unfinished work: "This early in the development cycle, when the thing has yet to work even once as an application I would trust with my data, it seems wrong to suddenly be worried about all the little details that a software release brings out. And yet working without interim deadlines can lead to a lot of aimless wandering through the land of coding possibilities."

At a developers' meeting a week before the release, Ducky Sherwood asked, "Are we going to have a party for 0.2?"

"A wake, perhaps," Toy answered.

"A moment of silence," Kapor suggested.

■ ▌ ▐

On October 2, the developers gathered for a postmortem on 0.2. What had gone wrong? What should they do differently next time around?

Morgen Sagen suggested that 0.2 really could have been labeled as just another interim milestone. "A dot release should have new features."

"We were half in one world, half in another when the clock stopped," Anderson said.

Brian Skinner spoke up. "Some features didn't make it in, but a ton of features did. And we have the repository now that we're going to live with."

One after another the developers offered a critique. Was it worthwhile to stop and fix bugs when so much of Chandler's architecture was still in flux? How could they even think of "fit and finish" when so much of Chandler wasn't even working? And did they have any prayer of reaching what was now their target date for a release of Chandler 1.0, Canoga, by December 2004, which was fifteen months away?

Toy registered each criticism with a bullet point on the whiteboard.

"One of the things we're saying is, we're really wired to believe that releases have goals," Toy said. "And 0.2 really had no goal. It had effects, but it didn't achieve anything. It was a nonevent. There was no party because there was nothing to celebrate."

OSAF would never again have a release quite as gloomy as 0.2, partly because it learned from its woes and partly because the organization would never again plan a release totally by the clock. Each future release would take shape around a set of features as goals, and its schedule would evolve (read: lengthen) as necessary. The experience didn't derail work on Chandler, but it did leave a kind of organizational scar.

One of the conclusions Kapor drew from 0.2—Brooks's Law be damned—was that OSAF was going to need more hires if it had any hope of reaching 1.0 in a reasonable amount of time. Several new faces already sat around the table at the 0.2 postmortem, and more would soon be on the way. Heikki Toivonen, a taciturn Finn who had been a longtime Mozilla Project contributor and, most recently, a Netscape employee, had come on board to focus on Chandler's security. Stuart Parmenter came from Netscape, too, where he'd been the resident wunderkind. While in high school he had become a volunteer contributor to Mozilla, not long after Netscape decided to release its browser code to the open source world. When he turned eighteen, he left his Georgia home, moved to Silicon Valley, and worked at Netscape until AOL shut down the browser team. At OSAF he started working with Hertzfeld on agents.

The other newcomer, a young woman named Mimi Yin, sat quietly, taking in the meeting's collective self-criticism. Her title was UI (user interface) designer. A student of music and dance, she had earned money as a Yale undergraduate by designing newsletters and Web sites for the student health service. When she graduated in 1999 at the height of the dot-com boom, she moved to the Bay Area to work for an e-commerce company and learned interface and information design on the job. During the post-bubble downturn, she split her time between choreography projects in New York and design jobs on the West Coast. Most recently she had been a contractor for Microsoft, working on a redesign of the MSN online service.

She had interviewed once with OSAF in March 2003, but nothing came of it. She was sharp and multidisciplinary but, some of her interviewers felt, maybe a little green for the job. In the fall, Chao Lam called her up to come in again. "I guess they haven't found the magical person they've been looking for," she thought.

■ ■ ■

Yin's arrival, coinciding with the start of the Chandler 0.3 cycle in October 2003, kicked the design process back into high gear. A new design group formed: Kapor, Hertzfeld, Chao Lam, Ducky Sherwood, Brian Skinner, and Yin. They began to meet three afternoons a week, taking new runs at some of the central ideas of Chandler—like the elimination of data silos. Yin's ability to whip up quick sketches on the whiteboard or the wiki to illustrate brainstorming ideas helped moor the discussions and elucidate hidden problems.

OSAF's flexible schedule allowed Yin to take dance classes most mornings, then come in for afternoon meetings—and post wiki pages anytime she wanted. She began thinking aloud in public, adding new pages to the wiki "Jungle" area at a furious pace, with CamelCase titles that broadcast their brainstorming intent:

```
BagsOfAttributes
PIMParadigmShift102003
HowCloselyTheUIShouldMapToTheDataModel
BalancingPointFoundWhatDoesItMeanToGetTheBasics
    ReallyRight
WhatKindOfMessHaveWeGottenOurselvesInto
```

Soon after starting work at OSAF, Yin attended a daylong seminar given by David Allen, a productivity coach whose book *Getting Things Done* was establishing near-cult status among programmers. Kapor knew Allen—they shared an avid interest in personal organization schemes—and Allen had already visited OSAF once to present a digest version of his principles (known as GTD to fans). But Yin was the person at OSAF who would take a systematic look at how the ideas in GTD might help shape Chandler.

Getting Things Done suggests that modern life and work leave us mentally beset by a host of incomplete tasks: physical objects lying around us waiting to be dealt with, incoming emails that need to be answered or filed, documents and publications that we're supposed to read, things we've promised other people we will do. This heap of "open loops" collectively constitutes what Allen calls our "stuff." GTD proposes that we can stop feeling overwhelmed by our stuff and take charge of it by creating a "trusted system"—on paper or digitally, it doesn't matter.

Into this system, we place our stuff so that we don't have to worry about it. With the system, we define concrete "next actions" for each of our projects. From the system, we can ultimately achieve what Allen, a martial arts devotee, calls "mind like water"—freedom from anxiety-inducing overload and that gnawing sense of "Oh, shit, what else have I forgotten to do?"

One central GTD principle calls for committing a certain amount of time regularly to processing your inbox—that is, considering each bit of stuff you have dumped into your system and deciding what to do with it next. If you can do what needs to be done in two minutes or less, Allen advises, just do it. Otherwise, decide if it's something to file, discard, defer, or classify as part of a particular project with a next action.

With Kapor's blessing and encouragement, Mimi Yin began to consider how Chandler might function as a GTD-style system for stuff processing. Kapor was clear that he didn't want Chandler to be too narrowly tied to GTD, but as the Design Group entered a season of intensive work in fall 2003, it periodically returned to Allen's ideas for inspiration. In particular, Yin emerged with the concept of a "dashboard" interface to help Chandler users process a torrent of incoming emails and other items without having to decide each one's ultimate fate. Yin's dashboard was a new kind of inbox that would present users with a top-to-bottom flow of items organized by past, present, and future. Older stuff would recede off the top of the screen into a storage area; deferred and future-scheduled items would get moved out of view at the bottom of the screen; and everything that needed to be processed would await the user's attention in the center. Yin's dashboard idea belonged to a tradition of time-line-oriented interfaces, like David Gelernter's Lifestreams project, that proposed chronology as the organizing principle for information management. But it mixed in Chandler's commitment to silo-leveling and David Allen's getting-things-done principles to come up with something unique.

OSAF itself was having its share of trouble getting things done, but if anyone there noticed the irony, the observation never made it into a blog post or wiki page. As Yin began the spadework of designing Chandler's user interface, one observation in *Getting Things Done* might have felt especially relevant: "Things rarely get stuck because of lack of time. They get stuck because the doing of them has not been defined." Yin's arrival furnished OSAF for the first time with someone whose full-time job was to ask and answer questions about Chandler's user interface and define it. On the one hand, this meant that even though the Design Group was not yet sharing its output with the programmers, the software developers finally felt that a fleshed-out design for them to work from might be coming soon. On the other hand, as Yin caught up with the previous work on Chandler (Brooks's Law in action on the design side) and began contributing big new ideas, the members of the Design Group itself found themselves back at first principles.

Over and over Chandler had returned to square one. There were the early brainstorming sessions in which Kapor and Morgen Sagen had fastened on the RDF approach. The interminable debates over how to design a flexible repository that would work well on a network. The extended era of document architecture explorations. Now, once again, OSAF's team was asking basic questions: How do people organize information? And how can an innovative software tool help them?

■ ■ ■

It was too much for Andy Hertzfeld, whose impatience with Chandler's lack of momentum finally came to a head. He had been feeling steadily worse about the project for some time. First of all, as he later explained to me, he had been finding it hard to work as a volunteer among employees— even working for someone he liked as much as Kapor. "They don't know whether to count on you or not. And because you're not getting paid, there's a lack of control." He also felt that Yin's approach was too abstract, too academic, and the two never clicked.

But the real problem was just plain speed. "I'm the kind of developer who likes to throw lightning rods around. To make a great program, there's got to be at least one person at the center who is breathing life into it. In a ferocious way. And that was lacking." At one point earlier in the year, frustrated by how slowly Chandler was moving, he had cranked out a basic address book "parcel" for Chandler: "I just designed it and wrote something. There wasn't all the discussion. My way of writing code is, you sculpt it, you get something as good as you can, and everything's subject to change, always, as you learn. But you climb this ladder of learning about your problem. Every problem's unique, so you have to learn about each problem, and you do something and get a better vantage point. And from that vantage point you can decide to throw it out. Code is cheap. But often it tells you what to do next."

But the address book never got integrated into the main tree of Chandler code. In fact, before too much longer, Kapor had decided to deemphasize Chandler's address book; someday the program might include one, but the feature now had low priority. So Hertzfeld decided to concentrate on fleshing out his idea for agents in Chandler while he participated in the summer's Document Architecture marathons.

In the fall, he left for a monthlong vacation in Hawaii with his wife. When he returned, the discussions the Design Group had begun left him flabbergasted. It seemed to him that they were ignoring all the work from the Document Architecture meetings—all the painstakingly defined vocabularies, all the careful sketches of blocks and screens—and were starting from scratch. Bafflingly, that seemed to be what Kapor wanted.

"When I told Mitch about it, he said, 'Well, I'm trying something different. We'll have to come back and realign things.' But wait a second! They're already building it! So I assumed that Mitch had thrown out CPIA. I was wrong, he didn't. But I just didn't get it. I think Mitch was a little unsatisfied with CPIA and how it turned out, even though he instigated it, so he wanted some fresh thinking. But, man, it made me feel like the thing was just circling. I still have infinite respect for Mitch, but he just lost me there."

As Hertzfeld withdrew some energy from OSAF in the latter part of 2003, he began a new project on his own: At a Web site called Folklore.org, he built a little software tool that borrowed aspects of both blogs and wikis to enable groups of people to contribute and share stories. Then he began posting tales from his days at Apple when he helped build the first Macintosh. The stories piled up, and the site grew. Then computer book publisher

Tim O'Reilly, whose company specialized in open source development manuals and who, incidentally, sat on OSAF's board, saw Folklore.org and proposed that Hertzfeld turn the tales into a book.

Hertzfeld agreed. He now had a six-month deadline to turn in a manuscript. Aside from a little advice and an occasional social visit, his work at OSAF was over.

■ ■ ■

Beginning in the summer of 2003, a steady flow of reports began emanating from Microsoft's Redmond, Washington, headquarters, with teasing glimpses into the world of the next version of Windows, code-named Longhorn. Longhorn, readers of the computer industry's distant-early-warning-system press learned, was going to be absolutely *incredible*. It would feature a stupendously beautiful and stunningly fast new graphics system called Avalon. It would provide a remarkably powerful and unprecedentedly versatile new networking system called Indigo. And at its heart would lie an astonishingly flexible and innovative new file system called WinFS.

In interviews, Bill Gates trumpeted Longhorn's possibilities. And the promises he was making for it sounded remarkably similar to the goals Kapor had set for Chandler, including the focus on breaking down the silos that lock different kinds of information away from one another. The WinFS file system would even allow you to store the same file in more than one place!

"Moving your files from this machine to this machine, getting your email, your calendar—it's painful," Gates told *USA Today*. "Say you have a work calendar and a family calendar. Is it really easy to coordinate your family's schedule and see which events should be on both ones? Longhorn makes it easy for your information to show up on any device. It makes it easy to navigate that information. . . . If I said to somebody today how on your PC do you keep track of stocks, movies, music, restaurants, you can do it, but it's pretty painful, pretty manual. The system doesn't have this innate understanding of all the things you deal with in typical life. You go and get directions on the computer, you get this funny Web page, and you probably just print the thing out. The idea of storing that, having it when you're offline—anyway a lot of these things are still pretty complex. So Longhorn is a change of the user interface to unify a lot of things that have been disparate."

Microsoft hadn't yet been foolish enough to specify the date on which Longhorn would be available. The company's record of delays in delivering promised upgrades to Windows was notoriously bad. But the Longhorn drumbeat was a reminder to Kapor and everyone at OSAF that the longer they took in delivering Chandler, the less chance their project would have of changing the world.

It had been a year since Kapor had suggested that Chandler 1.0 would be ready by the end of 2003, and now even meeting an end-of-2004 dead-line looked tough. He began working intensively with Chao Lam to try to prioritize Chandler's lengthy feature list. Chandler had promised to tackle a lot of broad areas—email, calendars, tasks, contacts, sharing, synchroniza-tion, search, security, agents—and each of those areas in turn had a long list of subfeatures. Chao Lam organized everything into a series of spreadsheets that were later moved onto the wiki, and he and Kapor began to decide what should go first and what should wait.

Kapor's attention turned more and more toward hiring. At the end of the summer he had posted to his blog: "Our head count has been fairly flat for some months at about 12. . . . Our plans call for a maximum size of over 30 in order to complete Canoga and Westwood in the next two and a half years." But finding the right developers wasn't easy, and integrating them into OSAF took time. "We can only absorb, on average, two or perhaps occasionally three new people per month without creating more chaos and disruption than I deem advisable," he wrote.

In October, Michael Toy presented Kapor with a surprise resignation. Some friends were starting a new software venture to streamline the business of call centers, allowing workers to take calls from their homes, and move it all to the Web. They had offered him the chance to be a vice president. Though he had thought of OSAF as a dream job, he was surprised to find himself drawn to the job offer.

Although Toy's self-disparaging comments had always seemed an endearing part of his personality—one week the task list he publicly posted read, "Continue in my more focused attempt to talk to people who I am allegedly helping by being their manager"—it turned out that he honestly felt he hadn't been doing a very good job. He explained his decision to me later.

"Mitch Kapor is spending millions of dollars a year and does not want to have responsibility for little details. He wants to find someone to make sure his money is being well spent. And that's a level of detail that I could not provide. And a level of assurance that I could not provide. Mitch was really patient in just saying, 'I will help you get better at these things.' So I have no complaints about the position I was in. But in the end, I was just being asked to provide a service that I wasn't very good at.

"And then I frankly admit that I am heavily biased toward: Let's ship something and put it in people's hands and learn from it quickly. That's way more fun, way more interesting, and, I think, a much better way to do things than to be sure that what we're doing is going to last for ten or fifteen years. It looks like Chandler is trying to be very architecturally sound and to be almost infinitely willing to delay the appearance of the thing in order to have good architecture. It's Mitch's money, so he can make that trade-off any way he wants. And it could be that that willingness to go slow is going to pay off hugely in the future. But it's really hard for someone who wants to ship software next week."

■ ■ ■

Toy's departure was orderly and on good terms—he spent most of November at OSAF tying up loose ends—but it left Kapor in a quandary. The last time around it had taken months to find the right development manager. This time, by coincidence, Kapor found a strong candidate immediately. He interviewed well and accepted an offer but then called back a couple of days later to say sorry, his spouse would not move to the Bay Area.

The prospect of a prolonged search for a manager, just as OSAF was beginning to get traction on Chandler's design and to get realistic about its priorities, troubled Kapor. They would keep searching, he announced, but they weren't going to let the empty chair impede their progress. It was becoming clear that if OSAF stuck to the original three-month, clock-driven release schedule and tried to finish Chandler 0.3 by the end of December 2003, it would be no more satisfying than 0.2. But the developers were making headway toward putting some of Anderson's new CPIA blocks into place, the design meetings were productive, and Kapor decided it was more important to produce an 0.3 that showed visible progress than to meet the clock-imposed deadline.

In Toy's last weeks at OSAF, Katie Parlante received new authority and a new title, Chandler software architect, matching John Anderson's. With Toy's departure, Kapor decided to structure OSAF's development effort into three groups: Design, Repository, and Apps (applications), where the bulk of OSAF's programmers would labor to build the Chandler front end. Parlante, who had quietly begun organizing the Apps Group before it had even been formally constituted, would run this core team.

As for overall management of the development effort, Kapor announced that while the search continued to replace Toy, he would take on the job himself.

"I will be focused on development details more than at any previous time in OSAF history," he told his staff. "Hopefully, that will be a value-added and not a value-subtracted activity."

■ ■ ■

By now, I know, any software developer reading this volume has likely thrown it across the room in despair, thinking, "Stop the madness! They're making every mistake in the book!"

A good number of those programmers, I imagine, are also thinking, "*I'd* never do that. I'd do better."

A handful of them might even be right.

After a year of sitting in on OSAF's work, I, too, found myself wondering, "When are they going to get going? How long are they going to take? What's the holdup?" Software time's entropic drag had kicked in, and nothing seemed able to accelerate it.

Couldn't the Chandler team see how far off the road they had driven? At every twist of the project's course, Kapor would freely admit that he would have proceeded differently if he had known before what he later learned. Then he would sigh, shrug, say, "Hindsight is always twenty-twenty," and get back to work.

In the annals of software history, Chandler's disappointing pace is not the exception but the norm. In this field, the record suggests that each driver finds a different way to run off the road, but sooner or later nearly all of them end up in a ditch.

If you are one of those programmers who are certain they could do better, ask yourself how many times on your last project you found yourself thinking, "Yes, I know that we probably *ought* to do this"—where "this" is any of the canon of best practices in the software field—"but it's a special situation. We're sort of a unique case." As Andy Hertzfeld put it, "There's no such thing as a typical software project. Every project is different."

In June 2004, *Linux Times* published an interview with Linus Torvalds, Linux's Benevolent Dictator.

"Do you have any advice for people starting to undertake large open source projects?" the interviewer began.

"Nobody should start to undertake a large project," Torvalds snapped. "You start with a small *trivial* project, and you should never expect it to get large. If you do, you'll just overdesign and generally think it is more important than it likely is at that stage. Or, worse, you might be scared away by the sheer size of the work you envision. So start small and think about the details. Don't think about some big picture and fancy design. If it doesn't solve some fairly immediate need, it's almost certainly overdesigned."

Torvalds didn't mention Chandler by name, but anyone familiar with the project who read his words couldn't help seeing the parallels, and anyone working on the project who heard them would doubtless have winced.

Yet even if you took Torvalds's advice—even if you started small, kept your ambitions in check, thought about details, and never, *ever* dreamed of the big picture—even then, Torvalds said, you shouldn't plan on making fast progress.

"Don't expect to get anywhere big in any kind of short time frame," he declared. "I've been doing Linux for thirteen years, and I expect to do it for quite some time still. If I had *expected* to do something that big, I'd never have started."

CHAPTER 7

DETAIL VIEW

[JANUARY—MAY 2004]

ard rain pelts the window of the big conference room on Howard Street. The shade is down; a projector is humming. It is the first ever "demo day" at OSAF, on a gloomy January morning in 2004, and all eyes are on Jed Burgess's desktop splashed across the room's back wall.

Burgess starts Chandler as John Anderson talks. It takes an agonizing full minute. Anderson's chatter fills the wait: "Our launches are not very zippy, you'll notice." Finally, windows begin to appear.

"This is the new repository viewer," Anderson says, as the screen fills with line after line of bracket-inflected XML data. "It's completely on top of our new architecture."

To a layman the barebones image wouldn't seem terribly impressive, but the observers in the room are rapt. They understand that this one little working component of Chandler—which retrieves a set of data from the repository and displays it, raw, to the user—has been assembled from OSAF's own parts, building-blocks from the "new world" they have painstakingly conceived and constructed over the past months.

"One of the coolest features about a CPIA block is that it's almost all data and very little code," Anderson says. "That's a pretty large departure from the way user interfaces were developed in the past. We're moving to a model where these UI elements are mostly data that you can inspect." In other words, everything the user sees on Chandler's screen—every window and button and menu—is represented as a piece of data in Chandler's own repository.

"So can you make these changes live, on the fly?" Mitch Kapor asks.

"Yes, you don't have to write code," Anderson answers. Just alter the data, and what you see on screen will change.

"Even at this level, that's a win," Kapor says.

Demo day seems to be working. Recognizing the malaise that had descended upon the team during the 0.2 release, Kapor decided to set aside time at their weekly meetings for the programmers to show their work to one another, walking the crowd through the small increments of progress they have recently made. As the two dozen attendees of the meeting applaud Anderson and Burgess, there's a sense of buoyancy in the room.

Next up, Morgen Sagen describes the new system he has assembled for automatically testing Chandler. Three computers now sit under his desk, one each for the Windows, Mac, and Linux versions of the program. Scripts running on these boxes constantly look for new "check-ins" of Chandler code; each time a developer checks in a change or addition to the source code repository, these computers test it to make sure that it hasn't "broken the build"— that the new changes, combined with all the existing Chandler code, haven't caused the resulting programming assemblage to fail one or more of Sagen's tests. This doesn't ensure that the changes are bug-free—far from it—but it's a first line of defense against program breakage and programming error.

The three sentinel computers send the output of their tests to a program called Tinderbox, which publishes the results to a Web page with three graphs that constantly display the current state of the build. Green means it passed the test; orange means one test failed; red means the entire build failed. The page is publicly accessible to developers and curious onlookers anywhere. (You can check the project's build status right now at http://builds.osafoundation.org/tinderbox/Chandler/status.html.)

The next demo is by Ted Leung, the latest addition to the OSAF development crew. Leung had worked at Apple on software for the Newton handheld device; later, at IBM, he had helped with some of the early work on the XML data standard. During the dot-com bust he had become a key participant in the Apache open source project. The original announcement of Chandler caught his eye, and he began following the project, showing up on OSAF's Internet Relay Chat (IRC) channel under his handle, "sprout." He lived on Bainbridge Island, across Puget Sound from Seattle; as long as OSAF was a "face-time" operation where employees were expected to meet and work in the same physical space, he figured a job was out of the question. But Kapor understood that "real" open source projects drew on contributions from far-flung coders coordinated across the Internet and decided it was time to change OSAF's approach. Leung already knew how to do "distributed development" from his Apache work; Kapor invited him to join OSAF and work from home, with a visit to the San Francisco mother ship every few months.

Today Leung is going to demo some tests he has rigged to see how well the Chandler repository can handle large quantities of data. A dedicated blogger, he devised a program that would load twenty megabytes of blog posts in RSS format into Chandler, creating eleven thousand Chandler items. There's only one problem: While a colleague at the OSAF office runs the demo on screen, Leung is patched in by phone, and the phone link is only half working: He can hear San Francisco, but San Francisco can't hear him. So he narrates the demo by typing his explanations into the IRC chat room, which is projected on the wall, where his colleagues can read them.

Kapor, at once frustrated by the balkiness of the phone technology and impressed by the ingenuity of the work-around, shakes his head slowly from side to side. "This has to be a first."

Andi Vajda is last on the day's demo roster. He shows a bit more of Leung's repository stress tests, then switches the subject to full-text indexing. He walks over to the whiteboard and begins to explain. If users are going to be able to search their Chandler data, the program will need to index every bit of text in its user's repository. As it happens, there's already a great open source text-search tool called Lucene, and Chandler could incorporate it. But there's a hitch: Lucene is written in Java; Chandler is based on Python. How to integrate them?

"I had to do a certain amount of acrobatics," Vajda says, a proud smile stealing onto his face. He goes on to describe them, leaving the nonprogrammers in the room far behind. At the end of a long chain of compiling and "wrapping" the code from Java through an open source compiler called gcj and another tool called SWIG (for Simplified Wrapper and Interface Generator), Lucene could happily work with Python and the Chandler repository. Vajda then demonstrates the code on a 2.5 megabyte text file— a 1992 CIA World Report. For the program to index the document's approximately 300,000 words, it takes all of 1.3 seconds—faster than Chandler, at that point, would take to do just about anything.

As Vajda walks his colleagues through the baroque transformations of code he has orchestrated, his eyes shine and his eyelids flutter. When he mentions that he stayed up to the wee hours on Christmas Eve to finish the project, it doesn't sound like a complaint; evidently he'd been having way too much fun.

At the end of the demos, Kapor gets up, visibly stoked. "Some of our infrastructure bets are beginning to pay off." He announces that they've removed the job posting for a new development manager. "I'm just a control freak, and I wasn't ready to give up the job," he jokes. But it's clear that the structure that emerged in the wake of Michael Toy's departure is working. Progress is taking place. Code is getting written. And here and there, in corners of the development team, you could even sense a little of that quality of urgency that Frederick Brooks called "hustle."

▌ ▍ ▐

OSAF released Chandler 0.3 on February 26, 2004. The new release incorporated a working version of CPIA and a more mature repository, but it remained essentially unusable—of interest only to outside software developers who might be curious about what Kapor and his team were up to. But the mood around OSAF was the opposite of 0.2: There was a T-shirt signing and a little chocolate cake celebration around the kitchen area of the office, and some OSAF old-timers, like Lou Montulli and Aleks Totic, showed up to join the cheers. The programmers felt they had some traction.

The progress also meant that the developers in the Apps Group, who had been content up to this point to let the design team work at its own pace, began to get itchy for a more thorough and final roadmap of how Chandler should look and behave. As the prospect came into view of actually building the program's real user interface—not a prototype or a stopgap but the real thing, using the real repository and the architecture that the Chandler team had laboriously invented over the past year—the programmers' hunger for information grew palpable.

They needed details. They needed blueprints. They needed specifications—a word so vital to the work of programming that over time it has shed all but its first syllable, becoming the terse, irreducible *specs*.

▌ ▍ ▐

If you've seen the brilliant heavy-metal parody movie *This Is Spinal Tap*, you might recall the scene in which the washed-up band tries to revive its career with a bombastic Stonehenge-themed number festooned with Celtic kitsch. The debut runs aground when the centerpiece descends from the ceiling to land in a circle of dancing midgets: The replica of a Stonehenge arch is a puny foot and a half high.

"I do not, for one, think that the problem was that the *band* was *down*," the lead singer complains to the manager in a hotel room postmortem. "I think the problem was a Stonehenge monument on the stage that was in danger of being *crushed by a dwarf*!"

The root of the problem? A bad spec. On the back-of-a-napkin sketch drawn by guitarist Nigel Tufnel at a coffee shop late one night, an extra apostrophe had transformed an eighteen-foot megalith into an eighteen-inch farce.

"It's my job to do what I'm asked to do by the creative element of the band," the manager protests. "Nigel gave me a drawing that said eighteen inches."

"But you're not as confused as him, are you?" the singer answers. "It's not your job to be as confused as Nigel."

When the manager confronts the set designer, she's indignant. She pulls the napkin out of her briefcase: "Look, look. This is what I was asked to do. Eighteen inches. Right here. It's *specified*!"

■ ▮ ▌

But it was in the spec! Or, *It wasn't in the spec!* These cries are the inevitable refrain of every software project that has hit a snag. Writing the spec, a document that lays out copiously detailed instructions for the programmer, is a necessary step in any software building enterprise where the ultimate user of the product is not the same person as the programmer. The spec translates requirements—the set of goals or desires the software developer's customers lay out—into detailed marching orders for the programmer to follow. For a personal finance software product, for instance, requirements would sound like this: *Must support ledgers for multiple credit card accounts. Needs to be able to download account information from banks.* The specs would actually spell out *how* the program fulfills these requirements. Visual specs exhaustively detail how each screen looks, where the buttons and menus are, typefaces and colors, and what happens when there is too much text to fit in a line. Interaction specs record how a program behaves in response to the user's every click, drag, and typed character.

The spec is the programmer's bible, and, typically, the programmer is a fundamentalist: The spec's word is law. Programmers are also, by nature and occupational demand, literal-minded. So the creation of specs calls for care and caution: You need to be careful what you wish for, as in a fairy tale. (Everlasting life? Don't forget to specify eternal youth as well!)

One of the oldest jokes in computing tells of the flummoxed user who calls a support line complaining that although the manual says, "Press any key to begin," he can't find the "any" key anywhere. The joke looks down its nose at the hapless ignoramus user, but the novice's misunderstanding actually mirrors the sort of picayune context-free readings that are the specialty of master programmers. Specs, as their name indicates, are supposed to bridge the realm of human confusion and machine precision through sheer exhaustive specificity. The programmer is to be left with no ambiguities, no guesswork. But the effort is inevitably imperfect. In part that's because the specs' authors are human, and the language they are written in is human. But it's also because the effort to produce a "perfect spec"—one that determined and specified every possible scenario of a program's usage and behavior—would prove an infinite labor; you'd never finish spec writing and start coding.

In practice, human impatience and the market's demands conspire to make underspecification the norm. And at OSAF the problem wasn't knowing when specs were finished but, rather, figuring out how to get them started. Despite the new feeling of momentum on the team, a few complex areas remained, in the terminology of OSAF old-timers, *snakes*.

Sharing information was to be one of Chandler's most important features, but every time the design team confronted the subject, it got stuck in a loop of unresolved questions or, in the language of project management, "open issues." How would one user initiate a share with another? What exactly could be shared? A single item? A whole collection of items? A particular view or screen layout of a collection? How would Chandler handle "permissions" when a user wanted to share some items but not others? If some shares allowed the sharee to make changes in the shared item and others didn't, how would the sharer make that choice? If two users share an item and they both alter the item when they're offline, how would Chandler resolve the conflict when they go back online and it tries to synchronize their changes? And then there was the thorniest sharing problem of all: "chain sharing." If I share an item with you, can you share it with a third person? Can she share it again? What happens when sharee number three or four or five makes a change in a shared item? Does it propagate all the way back down the sharing line? Another name for this problem was "n-way sync"—synchronizing some unknown (n) number of versions of a piece of data.

What started as a simple requirement—"Users should be able to share information easily"—very quickly mushroomed into a cloud of difficulties. For a while it worked to just say, "Let's defer those discussions," but everyone knew that eventually they would have to be hashed out.

Sharing wasn't the only trouble spot. There remained a host of open issues surrounding Chandler's approach to collections, or groups of items. In some ways collections were meant to be like the familiar folders of the desktop interface—containers for groups of things—but in other ways they were going to be novel. Items could exist simultaneously in many collections. Some collections would be rule-based. You'd simply tell Chandler to create a collection that, say, contained all items in which your business partner's name appeared, and the program would automatically keep it up to date. In these ways, Chandler's approach to managing information would draw inspiration from Apple's iTunes software for managing digital music collections, which offered its own version of these features. But Kapor also wanted to make sure that Chandler would allow for arbitrary exclusion and inclusion of items. In his favorite example, he might want a rule-based music collection to include any track by Bob Marley—but to leave out one particular live recording that he detests. iTunes won't let you do that.

"The first time around," Kapor explained during one discussion, "when I was designing Agenda, when I said we would want to have collections defined by rules but also have situations where you just wanted to put something in a group arbitrarily, I got enormous pushback. It was like I was violating some deep taboo of computer science. And it's true, there are a lot of very squirrelly cases. But we can't leave out features just because we think people will be afraid of them. Then you never invent the bungee cord."

But in order for Chandler to be able to handle Kapor's Bob Marley Exclusion Case, OSAF's design team had a lot to figure out: How exactly would the user create such collections and specify the inclusions and exclusions? How would Chandler store them? How could you make sure that people didn't get confused about the difference between excluding an item from a collection and deleting the item entirely from Chandler? How would all these different types of collections show up in the Chandler interface?

Then there was the problem that came to be known as "item mutability" or "polymorphism." Chandler's destiny was to level the silos that other personal information management systems built to segregate one kind of information from another. Emails and calendar events and to-do tasks and notes would all swim in a big informational pool that users could dip into as needed. But how exactly would a user go about taking, say, an incoming email about a party and turning it into an appointment on the calendar— or transforming an email from a colleague into a task on a to-do list? Would the email disappear or remain in your email collection? Would you end up with two items or one?

As the members of the design team struggled to get their heads and hands around these questions, Kapor reminded them of a programming maxim (widely attributed to personal computing pioneer Alan Kay): Simple things should be simple; complex things should be possible. Kapor wanted Chandler to be simple to use; he also wanted it to solve some famously hard software problems. Now he and his crew were hip-deep in the mire of trade-offs, trying to make it all work.

One day in April 2004, Chao Lam sent Mimi Yin a link to an article that he had found in a blog posting by a writer named Clay Shirky, a veteran commentator on the dynamics of online communication. Shirky had written about his rediscovery of an old article by Christopher Alexander, the philosopher-architect whose concept of "patterns" had inspired ferment in the programming world. The 1965 article titled "A City Is Not a Tree" analyzed the failings of planned communities by observing that typically they have been designed as "tree structures." "Whenever we have a tree structure, it means that within this structure no piece of any unit is ever connected to other units, except through the medium of that unit as a whole. The enormity of this restriction is difficult to grasp. It is a little as though the members of a family were not free to make friends outside the family, except when the family as a whole made a friendship."

Real cities that have grown organically—and real structures of human relationships—are instead laid out as "semi-lattices," in Alexander's terminology. A semi-lattice is a looser structure than a tree; it is still hierarchical to a degree but allows subsets to overlap. Why do architectural designs and planned communities always end up as "tree structures"? Alexander suggests that the semi-lattice is more complex and harder to visualize and that we inevitably gravitate toward the more easily graspable tree. But this "mania every simpleminded person has for putting things with the same name into the same basket" results in urban designs that are artificially constrained and deadening. "When we think in terms of trees, we are trading the humanity and richness of the living city for a conceptual simplicity which benefits only designers, planners, administrators, and developers. Every time a piece of a city is torn out, and a tree made to replace the semi-lattice that was there before, the city takes a further step toward dissociation."

Yin read the Alexander piece—the path it took to reach her, via Clay Shirky's blog and Chao Lam's email, seemed to illustrate the article's point about complex interconnection—and immediately applied its ideas to Chandler's design. She emerged with an elaborate design for a Chandler browser that was loosely inspired by iTunes; it would allow users to navigate the items in their repository by mixing and matching (or "slicing and dicing") from three columns of choices (for instance, you might have a first column with only "emails" checked, a second with a set of selected names, and a third with a date range).

The browser design occasioned lengthy discussion but did not win any kind of fast consensus. In trying to imagine a different, less hierarchical structure for Chandler's information, Yin was smacking headfirst into the reality that hierarchies are embedded deep in the nature of software and in the thinking of its creators. A city may not be a tree, as Alexander said, but nearly every computer program today really *is* a tree—a hierarchical structure of lines of code. You find trees everywhere in the software world—from the ubiquitous folder trees found in the left-hand pane of so many program's user interfaces to the deep organization of file systems and databases to the very system by which developers manage the code they write.

"The tree" is the informal name for the directory of source code where developers check in their work. If a group needs to work on a part of the program on their own without disrupting the work of their colleagues, they form a new "branch" of the tree, separate from the "trunk." Later, when they want to reintegrate their work with the rest of the code, the new branch is merged with the trunk. When a testing system like OSAF's new Tinderbox reports that all is well with the project, the programmers say with some relief that "the tree is green."

So getting away from trees might not be so simple. In any case, Yin's browser also felt disconnected from the immediate problems and needs of the apps team developers. They were miles away from being able to implement anything like it. One of the next big hurdles they faced was implementing Chandler's "detail view." The Chandler screen was to be organized into four basic areas. Across the top a traditional area for chrome (menus, toolbars, and such). At the left a sidebar for listing and organizing collections of items. The center, the main area, would split into two separate parts: a summary table (the "big table" that had been a stumbling block through much of 2003) and a detail view. The summary would list items (emails, calendar events, anything the user put into Chandler); when the user highlighted an item in the summary, the detail view would show everything about that item.

This structure was familiar enough from traditional PIMs and email programs, but there was still a ton of work in figuring out exactly how to design and build each element. The summary view, though far from complete, was moving along. The detail view now loomed as the biggest hurdle. It wasn't just a matter of cataloging and listing all the text associated with an item; Chandler's silo-leveling ambitions meant that the developers faced some novel questions as well. The simple act of sketching what this view would contain and show, and what it wouldn't, became a focal point for decisions—what Kapor liked to call a forcing function. The job of creating the detail view, turning those choices into a functioning piece of code, fell on the shoulders of OSAF's newest programmer, Donn Denman.

Some programmers walk around in a dark cloud, as if every single bug they've ever had to fix has left a scar on their psyches. Others look like there's nothing they would rather be doing than hacking out code. Donn Denman is the second kind of programmer.

"When I was young, I loved puzzles," Denman told me when I asked why he had gone into his line of work. "I really had a passion for puzzles, like stringing bead puzzles. I can spend hours even now with a puzzle if it's well designed—trying to figure out how to get the ring free from the tangle, figuring out how it works even though it seems impossible.

"I also liked mathematics in school. I think those sorts of things appealed to me because when I got it right, I *knew* I got it right. In mathematics you can often answer the problem and then work it backward to see if it's the correct answer.

"The same is true with a program that works if it doesn't have bugs— you can see your success—whereas so many other things in life are subjective, and you don't know if you were successful or not. That appealed to my insecurity at the time. Also, I really enjoy being able to create new things, building things, taking things apart, seeing how they work, and fixing things. I think the objective reward is less important to me now. I'm older, I have more self-confidence, I don't need that feedback. But being able to create stuff, see things happen, build things—that's still a lot of fun."

Denman grew up in Yellow Springs, Ohio, home of Antioch College, where his father was a professor and which he attended. He had gone to work for Apple right out of college in 1979, in a cubicle right across from Andy Hertzfeld. At the young company he found a rebel spirit that reminded him of Antioch's countercultural ethos. Denman's obsession has always been with creating systems that allow people with little or no programming experience to automate their work on a computer. In that era, Basic was the tool for such work. At Apple he worked on porting a version of Basic from the Apple II to the new Apple III.

Then Hertzfeld recruited him for the Macintosh team—where his job was, again, to write Basic for the new computer, with an approach that took advantage of the Mac's revolutionary graphic interface. "We had quite a bit of time," Denman recalled, "and I had these ideas of how to build a new Basic from the bottom up. I was making progress but unable to really put limits on the project and manage it to get it done on time and ship with the initial Mac."

Denman kept working on MacBasic, but before it was ready, Microsoft had released its own Macintosh-based Basic—one that the Apple programmers felt was inferior and poorly integrated with the Mac's new design. Meanwhile, Apple's deal with Microsoft to license the Basic that ran on the Apple II was up for renewal. In return for a new license, which Bill Gates knew Apple badly wanted, Microsoft demanded that Apple shut down Denman's MacBasic. Apple sold the code to Microsoft for a dollar.

As Andy Hertzfeld tells the story on his Folklore.org site, "When Donn found out that MacBasic had been canceled, he was heartbroken. His manager told him, 'It's been put on hold indefinitely,' and instructed him to destroy the source code and all copies, but refused to answer Donn's questions about what was going on. Later that day Donn went for a wild ride on his motorcycle and crashed it, returning home scraped up but with no real damage, except to his already battered ego."

"I had put so much time and effort in, it was like my baby," Denman recalled to me. "My two-year-old had been taken from me. I think it took me ten or twenty years to get over that." Ultimately, he stayed at Apple and worked on AppleScript, another system for nonprogrammers. He finally left the company during its mid-nineties tailspin and rode out the tech industry bubble working on software for the digital cable TV delivery systems known as set-top boxes. The bubble's gyrations didn't make him rich but left him with enough money to take some time off. He enjoyed tinkering with his own projects on his own time, but when Hertzfeld started telling him about Chandler, his eyes lit up. He had been itching to return to his passion: scripting systems for the programming challenged. Kapor's project sounded like just the ticket.

Denman joined the Apps Group in April 2004 and sat down with Katie Parlante to figure out what to work on. The detail view assignment seemed like a good way for him to dive into the Chandler code and learn the idiosyncrasies of CPIA. He would be the first programmer to make serious use of CPIA who hadn't had a hand in inventing it, and the questions he asked would help refine it.

But in order to build a working first draft of the detail view, Denman needed a spec to work from. And there didn't seem to be one. Mimi Yin had a pile of wiki pages and rough "wire-frame" sketches of aspects of the detail view, but they were in constant evolution as the design team's meetings kept throwing up new questions about sharing and collections and item mutability. Denman realized he would have to confer with the designers and start writing his own specs. So in May, Denman and Yin, along with Chao Lam, began to meet almost daily in one of Howard Street's little meeting rooms—each a windowless space just big enough for three or four people, a table, and a wall-to-wall whiteboard. The rooms would heat up after an hour, and the dry-marker fumes would get stifling, but the group was determined to motor through the open issues and clear a way for Denman to start coding.

Yin explained the latest thinking on items. Most PIM programs required users to decide up front, when they created a new item, what it was: Were you creating a new email? A calendar event? A to-do task? Chandler would instead let you sit down, start typing a note, and decide later what kind of item it was. Like the human body's undifferentiated stem cells, notes would begin life with the potential to grow in different directions. This design aimed to liberate the basic act of entering information into the program from the imprisoning silos. It also made room for Yin's proposed solution to the item mutability problem: The mechanism users would employ to specify the "kind-ness" of an item would be called *stamping*.

Say you had typed a note—a couple of sentences about a meeting—and then wanted to put the meeting on the calendar. You would stamp the note as an event. Chandler's detail view would add fields that let you specify a date and a time; your generic note was now an event. Later, if you wanted to invite a colleague to the meeting, you could take the same note and stamp it as an email. A "to" field and a subject line would appear in the detail view. You would fill it out and click on a "send" button.

Stamping was a nifty idea, but it raised almost as many questions as it answered. Could you unstamp a note? And if you did, what happened to the information associated with the kind-ness you had removed (like the dates on a calendar event)? When you stamped a note and then looked at the "all" collection (a list of all the items of every kind in your Chandler repository), did you see one copy or two?

Yin started producing detailed sketches, and Denman looked them over, peppering her with questions: What does this label mean? How does this feature work? The designer had a lot of explaining to do.

■ ▪ ▪

It is a May afternoon, and Denman is struggling to finalize a spec for the first run at the detail view. He and Yin begin filling up the last bare patch in the corner of a whiteboard.

"I need some way to show when an item is new, and I haven't figured it out yet," Yin says.

"What is this?" Denman points to an icon of a little head that sits near the top of a detail view drawing. It's for sharing, Yin explains. It's what a user would click on to send an item to someone else. She wonders aloud whether the icon would belong on an email item at all, which would have its own controls for sending. "But if the email was turned into a task, then you'd want it." Denman scratches in his notebook. Yin pauses, turns away from the board, and says, "These are the things that give me nightmares."

"Well"—Denman looks up—"I'm going to help you slog through it. We'll build a subset of it all and see what works. So for 0.4, at least, let's say we don't have the little heads at all."

"No," Yin replies. "You need the heads; otherwise, there's no sharing."

"So let's say there's no sharing." Denman smiles. He might be joking; he might not.

Yin copies all the whiteboard jottings into an email to record their discussion and decisions. It's a reminder of how email has become the de facto warehouse for all kinds of information for so many computer users—and of how badly most email programs perform that role, with their lousy search capabilities and brain-dead management of attached files. As the OSAF developers kept saying, "We could really use Chandler for this."

"I'm going to take one more week to plan, and then hopefully I'll begin the implementation phase," Denman says. He adds, "I'm still having fun in my new job!"—with a half smile that suggests the fun is decidedly a half-glass-full-or-empty sort of thing, and he's actively choosing the full view.

"The feeling here is a lot like Apple in the early days," Denman says to me one afternoon during the stamping work. "That's one of the reasons I'm excited to be working here. It really does feel like we could change the world—though that's not what we're saying here, and at Apple, of course, we were saying that all the time!"

■ ■ ■

Stamping aimed to introduce a kind of productive ambiguity to the computer desktop that more closely mirrored the way people think. It was not a simple concept even for the designers who had invented it; for the developers who had to make it work, it was even trickier. Computer programs used silos and trees and similar unambiguous structures because they helped keep data organized and limited confusion. If an item belonged to one group, it did not belong to another; if it lived on one branch of a tree, it did not live on another.

Human language is more forgiving: One word can mean more than one thing. This flexibility provides a deep well of nuance and beauty; it is a foundation of poetry. But it leads only to trouble when you are trying to build software. As OSAF's developers struggled to transform the innovations in Chandler, such as stamping, from sketch to functioning code, they repeatedly found themselves tripped up by ambiguity. Over and over they would end up using the same words to describe different things.

Take *item*. To the designers an item in Chandler was anything a user thought of as a basic piece of data. A single email. An event on a calendar. A task or a note. But the back-end world of the Chandler repository also had items, and its items were subtly but substantially different from the front end's items. A repository item was a single piece of information stored in Chandler's database, and in fact you needed many of these repository items to present a user of Chandler with a single user item like an email: Each attribute of the user item—the subject line, the date sent, the sender's address, and so on—was a separate repository item. At different times in Chandler's evolution, proposals arose to resolve this problem—to "disambiguate" the word *item*. Maybe the term *user item* could always be capitalized. (This helped in written material, when people remembered to do it, but not in conversation.) Maybe another term for one or the other type of item could be adopted. (But some of those proposed, like *thing*, were even more ambiguous, and none of the proposals stuck.)

The Chandler universe was rife with this sort of word overlap. The design team kept using the term *data model* to refer to the list of definitions of data types that the user would encounter along with all the attributes associated with that data type. For example, the data model would specify that every note had a "date created," an "author," and a "body text." But to the developers, *data model* referred to a different, more technical set of definitions that they used to store and retrieve data at the level of their code. What the design team called the data model was really, in the developers' vocabulary, the *content model*.

Then there was the problem with the term *scheduled task*. In the design world a scheduled task meant an item on a user's to-do list that had a date and time assigned to it. But for the developers a scheduled task was something that Chandler itself had been told to perform at a particular time, such as download email or check for changes in shared information. Or consider the term *event notification*. For the designers this meant things like telling the user that new mail had arrived; for the developers an event was some change in the status of a code object, like a window closing or a click on a menu item, and notification meant sending news of that change to other code objects.

Kapor would observe these little linguistic train wrecks and shudder. "We need to speak one language," he would say. "We should all speak Chandlerese. We have to fix the vocabulary."

Finally, Brian Skinner stepped forward to try to do just that. Skinner had joined OSAF as a volunteer and helped Andi Vajda and Katie Parlante sort out the subtleties of the data model back when Vajda was just trying to get the repository started. Now a full-time OSAF programmer, Skinner had a knack for explaining developer-speak to the designers and design-talk to the developers. When the groups talked past each other, he was often the one to sort out the language. Why not, he proposed, set up a Chandler glossary on the wiki? It would provide a single, authoritative, but easily amended reference point for all the terminology floating around the project. It could literally get everyone on the same page.

Skinner took up arms against the sea of ambiguity. He produced dozens of glossary pages. He built a system for linking to them from the rest of the wiki.

It was a heroic effort, but it didn't seem to make much difference. For one thing, usage of the terms continued to change faster than his wiki editing could keep up. More important, the developers, who were already drowning in emails and bug reports and wiki pages, didn't seem to pay much attention to the glossary, and the pages languished, mostly unread.

The glossary's futility might have been foreshadowed by the outcome of another naming effort in which Skinner participated. OSAF's Howard Street headquarters had a half-dozen conference rooms, and Kapor decided that it would be useful for them to have names, so that instead of saying, "Let's meet in the little conference room around the corner from Donn's desk," you could just say the room's name. OSAF held a contest and solicited proposals from the staff. Skinner suggested using names from imaginary places; he won the contest and ponied up his $150 prize to fund a happy hour for his colleagues at Kate O'Brien's, the Irish bar down the street.

The fanciful names—the two main conference rooms were Avalon and Arcadia—captured the spirit at OSAF, where imagining new worlds was on the collective to-do list. It was only later that everyone realized what a bad idea it was to have the names of the most frequently used rooms start with the same letter: No one could ever remember which was which.

■ ▮ ▯

In a cartoon that you can find on more than one programmer's Web site— and, I would bet, on the walls of cubicles in software companies everywhere—Dilbert, the world's favorite downtrodden geek, says to his supervisor (the celebrated Pointy-Haired Boss, or PHB), "We still have too many software faults. We'll miss our ship date." The PHB replies, "Move the list of faults to the 'future development' column and ship it." In the final window, the PHB, pleased with himself, thinks: "Ninety percent of this job is figuring out what to call stuff."

The names programmers adopt for their abstractions are, as geeks like to say, "nontrivial." In software, labels and names matter; they are the handles by which you grab things.

Infants expanding their vocabulary beyond "Mama" and "Dada" learn by associating sounds with visible physical objects; we point to something and repeat its name. Teaching children the meaning of more abstract words is a lot harder; we try to explain one word in terms of another, or we tell a story.

Beyond the windows and text of a user interface, most elements of software are incorporeal and invisible. There is nothing to point to. So talking about them is unexpectedly difficult. This is one reason the whiteboard is such an iconic presence in any space where software is labored over; it provides a canvas for laying out the abstract processes of a complex program in ways that allow people to point to what they're talking about.

One of the first tasks a programmer faces with a new program is choosing labels for the variables and objects the program will use. A variable is a program's way of setting aside a spot in the computer's memory where it can put some information. Later, it will be able to retrieve that information, the variable's value, by addressing it—by calling out its name.

Machine memory demands a one-to-one correspondence between variable name and memory address. Human memory, by contrast, is richly layered and overlapping; its density and boundarylessness fuels hunches and sparks intuition. But this feature of the brain becomes a bug in the digital machine. If, thanks to errors or bad programming practices, two programs try to use the same location in the computer's memory to store a variable, unpredictable and unstable results are guaranteed (the computer will usually freeze or crash). Software abhors ambiguity.

Human beings remain far more capable than computers of reading clues from contexts and teasing out the most likely meaning of an ambiguous signal. But if we expect computers to understand us, we typically have to sort out any ambiguities beforehand. Don't ask a computer, even one adept at handwriting recognition, to distinguish between a scribbled letter "O" and a zero. (Old-fashioned computer terminals used to print out zeros with a slash through them to help make the distinction.) The word *disambiguate* did not exist before the culture of computing needed to invent it.

When programmers sit down to perform the almost biblical act of naming all of a program's variables, they must therefore aim to keep meanings clear and discrete. Many developers in the "let's suppose that . . ." stage of rough prototyping begin with the all-purpose stand-in names *foo* and *bar*. The origin of these labels is in dispute, though most likely they derive from the old army acronym FUBAR (Fucked Up—or Fouled Up—Beyond All Recognition/Repair/Reason/Redemption). Sooner or later, though, these meaningless placeholders must give way to real variable names that communicate things, like AccountHolderID or LastDateModified.

Carefully chosen names avoid the confusion of "namespace clashes" or "collisions"—the use of a term that means one thing in one context but something else in another. Wiki inventor Ward Cunningham even suggests that programmers keep a thesaurus on their desks. How you choose names matters, he argued in a 2004 talk, "because *people read* these names. And their first guess is that, you know, the object is what it says it is. But usually it isn't quite what it says it is." Names that meant one thing when the programmer's work began end up meaning something different once a thousand bugs have been fixed. Cunningham urged programmers to go back and fix their names—but admitted that most don't bother.

The quest for rigor in the naming of variables and software objects is almost as old as computer programming itself. In the early days, there was so little space available in the computer's memory that good programmers learned to conserve it by abbreviating their variable names. It really made a difference to name something *memloc* rather than *memory location*, particularly when a variable's name might be repeated hundreds of times in a program's code.

The most famous, and infamous, scheme for naming variables is Hungarian notation. It takes its name from the native language of its inventor, Charles Simonyi, but also from the often cryptic and unpronounceable labels it produces, which Simonyi's colleagues at Microsoft thought looked vaguely foreign. In Hungarian notation, the programmer appends a prefix to every variable name that gives anyone reading the code important clues about what sort of variable it is. Simonyi, who had developed the first ever graphic-interface word processor as a researcher at Xerox, became the father of Microsoft Word, now the world's most widely used writing software, and his Hungarian notation convention became the house style at Microsoft. Over the years, Hungarian's defenders have argued that it helps make code readable, but its detractors have maintained that it turns code into gobbledygook. Alec Flett, a Netscape programmer who later joined OSAF, contributed this parody to a mailing list:

```
prepBut nI vrbLike adjHungarian! qWhat's artThe
adjBig nProblem?
```

Given that the point of the naming convention was to add clarity to the act of writing code, it seems cruelly ironic that it suffered from a critical ambiguity originating in the first paper that Simonyi wrote to explain it. Hungarian notation, it turns out, *means two different things*. The Hungarian the programmers who built Microsoft Windows used, which came to be known as Systems Hungarian, was not the same Hungarian Simonyi had originally promulgated to the programmers of Word, which sticklers today call Apps Hungarian. When Simonyi said variables should be labeled according to "type," the Systems Hungarian people interpreted him to mean the same sort of "type" that most programming languages required you to choose for your variable (integer, floating-point, string). But Apps Hungarian, *true* Hungarian, meant "type" in a looser, more everyday sense, encouraging you to communicate more information about each variable that might actually be useful to a programmer reading the code.

Somewhere along the line, Microsoft's zealous program managers had misunderstood, or ignored or dismissed, Simonyi's intent. As a result, Hungarian had—to borrow the word open source programmers apply to a project that has suffered a schism—forked. So you could be coding in Hungarian and I could be coding in Hungarian and our code might never connect.

■ ■ ■

Communicating abstractions unambiguously—from programmer to machine, from programmer to programmer, and from program to user—is the single most challenging demand of software development. How can programmers and nonprogrammers avoid a Tower-of-Babel fiasco? There are some writers of code who have so finely honed their ability to talk directly to the machine in its unforgivingly precise syntax that they have lost the simple, put-yourself-in-the-other-person's-shoes empathy that keeps human conversation on track. If you were to send such programmers to a room with their colleagues and tell them not to come out until they have agreed on a vocabulary, you might never see them again.

OSAF's developers were typically on the more socially skilled end of the programmer spectrum, and still they constantly tripped over the "same word, different meaning" problem. If you're really lucky, you might get a developer or two with imagination, a patient temperament, and a willingness to listen who might find a kindred spirit on the product development or design side, and together they might clear away enough of the ambiguous-terminology underbrush to bare some common ground and resolve some hard questions. At OSAF something like that happened as Donn Denman and Mimi Yin huddled to hash out the minutiae of the detail view spec, talking over each little icon and interaction until they got to the point where Denman could begin coding.

■ ■ ■

One day at the end of May, Denman pulled Yin over to his desk and pointed to an arrangement of tinted rectangles on his screen. Strip out the text labels, and you might have been looking at a Mondrian. They started discussing how to shape the interface for a feature called "item conversations," which would allow users to append brief notes to a shared item. Then Yin stopped and stared at Denman's screen. "Wait a second. You mean this is live?"

"Yeah, it's in Chandler," Denman answered.

"Cool! I thought it was just an image."

"No. I love seeing stuff up here! First I get a general layout and then I start wiring things up so that the fields are real and the buttons start to do things."

A week later Denman showed the same screen to the entire OSAF staff at the Thursday meeting. "These are placeholders," he said. "This isn't really a demo of stuff that's built; it's a demo of stuff that's to be built. For a while we'll have this mishmash of things that work but don't look right, or things that look right but don't work."

For the next month Denman worked with Andi Vajda and Ted Leung, the repository programmers, and figured out how to expand the repository's capabilities to make stamping possible. Yin and the design team had focused most of their work on what they saw as the most likely cases of stamping: An incoming email gets stamped as a calendar event or a task; a plain note gets stamped as an email, a calendar event, or a task. But programmers are always thinking about "edge cases," and, inevitably, the system Denman and his colleagues built to implement stamping ended up as a generalized solution—one intended to allow you to stamp an item of any kind as an item of any other kind. They devised a new concept called the "mix-in kind." A mix-in kind defined a set of attributes associated with a kind; for a task they might include "Priority" and "Done Status." If you had an item that was an email and you stamped it as a task, the repository would now simply add this new set of attributes to those the email already had (like "sender" and "subject line"). There were, inevitably, "namespace collisions": The simple attribute "date" could mean one thing in the context of an email (date received) and another in the context of a task (date due or date completed). If your boss sends you an email about a meeting with a client, and you stamp that as a calendar event, the "who" for the email (your boss) is different from the "who" for the event (the client). Chandler's back end could disambiguate those dates and whos, but the program would still have to pick which one to display to the user. Denman's approach didn't solve all these problems, just enough of them to get something running.

In early July, Denman and a few Apps Group developers gathered around the projector in Arcadia (or was it Avalon?) for a code review of Denman's stamping work. The term *code review* can mean anything from an informal monitor-side chat to a weeks-long bureaucratic gantlet involving multiple layers of code inspection. OSAF had always taken a fairly loose approach, but the developers had recently all agreed to aim for more code reviews; stamping was sufficiently novel and complex to make it seem like a good choice for the attention.

Denman outlined the concept of "mix-ins" and reviewed some of the tricky cases. "When we receive an email, the date that's important is the date it's received. When we stamp it as an event, the date that's important is the date of the event. So I've put in a hack to deal with this." He scrolled down the screen, and the projector displayed new chunks of code. "I'm sorry this is so tortuous." More code rolled by. "Self's class has changed. I threw in a repository commit [storing the data's state] here so we'd remember we have this new kind. . . . If I can figure out what my new kind is, I assign it. Otherwise, I use Andi's mix-ins." The next chunk of Python on screen defined NewStampedKind. "This is crufty hack code that basically looks to see if there's already a kind that matches the bunch of attributes that are now assembled after the mix-in."

I couldn't pretend to follow everything Denman was saying. Sitting in on a code review as an outsider is like eavesdropping on specialist shoptalk of any stripe; you can't always tell what the crux of the matter is. Still, I understood that Denman's code had tackled the stamping problem by making Chandler a little nimbler; it could now identify an item's kind simply by looking at its set of attributes, and it could transform the item into a new kind without losing any of the attribute data from the item's previous existence.

I also understood that Denman felt the result still needed a lot of work.

John Anderson looked up as Denman finished. "I need some time to let it roll around in my head. It's great, but it seems complicated. My intuition says we could get around this."

The approach was complex, but it had one overriding virtue: One week later—about ten weeks from the time Denman had started at OSAF—stamping was working. It might only have been a first pass and might have had some snags, but it was up and running. Denman demoed it at the next Thursday staff meeting.

"It's pretty simple and straightforward," he said, clicking around his still new detail view. "Now we can change the kind of an item when we stamp it." He displayed a note in Chandler labeled Fantastic Recipe. "This starts out as a note. You can decide to make it a task by pressing the little task thing. And you can see that some of the attributes have now changed, it now has requester and requestee fields. There are three things we can stamp now: a task, a calendar item, and a mail object—and you can have combinations of these things. In theory, when new kinds of content items show up, you can stamp them, too. You could stamp a photo with 'mail-ness,' it will get mail attributes, and you can mail it.

"This is a fundamental concept—that things can change in kind. I'm not doing most of the work. It's Andi in the repository. The item doesn't change its ID. It's the same object, but it has evolved. When you change a kind, it goes and looks around to find if there are attributes that match the kind. If it finds them, it applies them. If it can't find anything that matches, it creates a new kind on the fly."

The demo ended in a burst of applause, and Kapor, smiling, spoke up. Aside from sharing, stamping had been the biggest snake in OSAF's road. If not slain outright, it was plainly well under control now. The silos were toppling! But they still had far to go.

"There are stamping-related features throughout the UI—we've implemented maybe one percent of them," Kapor said. "When we pull this off, it will be one of the really distinctive features from all other PIMs. Things will come in, you'll click on them, and they'll show up in the right place. It will be a different level, and people will say, of course, that's how it should work! Why doesn't everything work that way? Our success will be measured by how well we reframe the question for people."

█ ▐ ▌

Recently, Kapor had reframed some more personal questions: How deeply was he committed to OSAF? And what job did he want there?

Since Michael Toy's departure in November 2003, he had been serving not only as keeper of the product vision and protector of the soul of

Agenda, but also as the hands-on manager of design and development. At the beginning, as his team broke out of the 0.2 doldrums and got more traction, he found the role invigorating. But as the months progressed, the volume of work kept expanding, particularly with new developers joining the project at a clip. Brian Kirsch, a young programmer who moonlighted as a musician, signed on to write Chandler's email code. David Surovell, an Apple veteran, would work on improving aspects of wxWidgets for Chandler, particularly on the Mac side. And Lisa Dusseault, a leader in the world of open source calendar standards, would share managing the growing crew of programmers with Katie Parlante.

The number of meetings at which Kapor's presence was required grew, and his audible exhalations at those meetings grew louder and more frequent. He would make a "T" for "time-out" sign with his hands. "Sorry. I'm experiencing buffer overflow," he would confess, and every geek in the room understood what he meant: *There's more coming at me than I can handle.* He wasn't the sort of boss to lose his temper; that set him apart from most other tech industry titans (Bill Gates's tantrums are the stuff of legend, and so are Steve Jobs's). In my three years at OSAF, I heard him raise his voice only a handful of times, always followed by a swift apology. But in the spring of 2004 his patience was plainly fraying.

At the same time his attention was being pulled in other directions by the revving up of the 2004 election cycle. Liberal hopes for defeating President George W. Bush were high—stoked by the nascent organizing and fund-raising power of emerging online communities, blogs, and organizations like MoveOn. Kapor liked to take the long view on prospects for "digital democracy," but he couldn't help getting caught up in the moment's spirit: He started a new political group blog called Of, By and For, and helped launch a new open source project for online community organizing called CivicSpace, whose software would live on, whatever the election's outcome.

For a long time Kapor, like many of his colleagues at OSAF, had made it a custom to spend a few minutes each day reviewing recent changes on the wiki. To see these you could visit a Web page, have a list emailed to you, or receive the information via an RSS feed. The system wasn't perfect—it would report an update whether someone had totally revised a page or just altered a comma—but it was a reasonable way to keep up with the expanding and increasingly far-flung Chandler team. In mid-May, though, as the pace of changes picked up and Donn Denman's work on the detail view filled the wiki with page after page of specification details, Kapor suddenly felt overwhelmed.

He had started OSAF partly because he believed in the ideals of open source, but also because, after his stint as a venture capitalist, he realized that designing software was his calling. Yet as Chandler's schedule grew more and more mired in software time, its completion date receding like a desert mirage, outsiders and insiders alike wondered how committed he was to keeping the project going.

When he announced OSAF in October 2002, Kapor had earmarked $5 million of his own money to support it. In 2003, the foundation received additional support from the Mellon Foundation and the Common Solutions Group, a coalition of university technology departments, totaling $2.75 million. With a growing roster (roughly two dozen by May 2004, and a plan to staff up to thirty), that money would last a while longer, but not forever.

Cash is typically the software industry's ultimate constraint, the limit that yanks projects out of software time's wormhole back to reality. But Kapor had plenty of it. If OSAF needed more, everyone assumed he would provide it—as long as he remained enthusiastic about Chandler. Esther Dyson, the technology pundit, had once joked that Kapor's company from the early nineties, On Technology, was "born with a silver spoon in its mouth." OSAF's spoon, perhaps, was merely silver-plated, but for an open source project, any kind of inheritance was unusual.

When people asked Kapor directly about his commitment to OSAF, he would answer, "This is what I want to do for the rest of my life." And he would point to the new long-term lease he and his wife had just signed expanding their combined operation to a second floor in the Howard Street building, providing room for OSAF to grow and space for other organizations they wanted to take under their wings.

But there was no question he had hit a crisis. The fun had drained out of his work: "It was like water turning to ice overnight," he later told his staff. "A phase change." He wouldn't quit, but he had to do something. He talked it over with his wife. She had an idea: Why not ask the rest of the managers at OSAF to redesign your job?

On May 18, right before the regular weekly Tuesday managers' meeting, Kapor sent around an email with the subject line "Additional issue":

> ```
> I love OSAF. I'm committed to its long-term
> success. We're taking another 10,000 feet on the
> 4th floor.
>     We're at an inflection point in the project.
> I can feel it. There's so much going on now with
> the addition of many productive new people to
> existing staff and a real focus on making a
> working product that my old methods of staying on
> top of things are now totally inadequate. New
> management methods are necessary.
>     . . . Now it's time to redesign my job. I need
> your help. There are obviously major things I've
> contributed but there are also ways I'm not the
> best person to do some of the management. I get
> in the way or am simply not bringing the best
> skill set. I want to do the right thing. I need
> some help. My feelings won't be hurt.
> ```

Kapor asked the group to meet without him and present him with a proposal. Several intense hours later they had a plan: Mitch should become more of a full-time leader and less of a hands-on manager. He should "empower" his managers, giving them more authority to make decisions on their own. They didn't need to go out and hire a single development manager, but there were other changes they wanted to make to improve the organization's structure. Parlante had already assumed a new mantle as "architecture coordinator" for big engineering decisions; Lisa Dusseault would now take on the detailed planning and scheduling of the development work.

Kapor was pleased. "We're in violent agreement," he told his colleagues when they presented him with their ideas. A week later he announced the plan to the whole staff, seated in his customary perch, a small rolled cushion for his bad back, his PowerBook casting a cool blue glow on his gray hair and beard.

"You should expect that the organizational structure is going to continue to evolve periodically," he told them, his eyes roaming the room. "The big challenge is, we have to constrain and mold our ambition level without losing the spark. The stuff I lose sleep over at night is, how are we doing on this? We want to make progress as fast as possible without exploiting human resources. How do you move faster without killing people? How do you build a community around the project without losing design integrity? And how do we get on an arc that moves from benign dictatorship to more small-group democracy?

"We've moved to more structure, not less, in order to do what we need to do now. Other than our transparency, it feels like a conventional software development project—I think, of necessity. How do we understand that this is a phase? So that over time we're not dug into a hole, but we actually have a project where there is greater involvement, greater self-determination on the part of the participants themselves? Putting it out there as an intention is useful so that when we make big decisions about what to do next, we have this in mind."

"Mitch has a big reputation," Andy Hertzfeld told me when I asked him why he thought Chandler was taking so long. "He doesn't want to get egg on his face. So we're a little more cautious than if it was a seventeen-year-old, where all he'd have is his code. We have a tendency to be very, very careful."

But hadn't Kapor proven himself? I asked. Eric Raymond had said that open source programmers were driven by a hunger for glory among their peers. Kapor had made his name many times over already. Shouldn't that make it easier for him to take risks?

"No, he still needs more glory," Hertzfeld responded. "We all need more glory as designers—to show we can design another great thing. Everybody who has a first success, especially when it's young, wonders: Was it luck, or was it skill? Well, it's a little of both. If you can do another really great one, it shows the world something."

■ ■ ▉

When observers expressed skepticism to him about Chandler's slow progress, Kapor would lift his owlish brows and comment, "I am nothing if not persistent." After all, he still heard regularly and frequently from people eager to ditch their Microsoft Outlook and Exchange programs and replace them with something better, more flexible and reliable. The very week his managers were redesigning his job for him, just before Memorial Day weekend, the Microsoft Exchange server that supported his wife's non-profit, the Level Playing Field Institute, had suffered a catastrophic melt-down. For three days the LPFI staff lost its email and schedules.

It felt as if Chandler was taking forever. But Kapor was determined that persistence would pay off.

CHAPTER 8

STICKIES ON A WHITEBOARD

[JUNE–OCTOBER 2004]

At Microsoft, as at many serious software companies, there has long been a corporate imperative to "eat your own dogfood." No, Bill Gates is not hazing his employees; the phrase means that software developers must themselves use the products they are in the process of building. (The origins of the usage extend back to olden TV times when Lorne Greene would hawk Alpo by exclaiming that the stuff was so great, he fed it to his own pooches.)

Here is how one Microsoft development manager once explained it in a blog posting: "Dogfooding . . . is something we do internally pretty aggressively to help chase out the last few production bugs in a server release. We've found that running your payroll, email, source control, and health benefits system on top of your own beta software can be a highly motivating way to drive quality into the product early in the development cycle. ;-)"

At OSAF, where dogs roamed the halls and a canine face festooned the flagship product, the term *dogfooding* felt just right; it had, as Mitch Kapor put it, "positive connotations." As work on Chandler plodded forward in the first half of 2004, Kapor and his team became steadily more convinced that the key to making better progress was to improve their software to the point where they could begin using it themselves—to make Chandler dogfood-able.

The term *dogfooding* may be popularly associated with Microsoft, but the practice is far older. At Xerox's legendary Palo Alto Research Center, where modern personal computing was invented in the 1970s, "We use what we build" was the motto of research team leader Bob Taylor. And that formulation was simply a variation on Doug Engelbart's bootstrapping. The goal of bootstrapping was to rev up a feedback loop so that today you would use the tool you invented yesterday to build a better one for tomorrow; dogfooding, by contrast, had the more modest and pragmatic aim of speeding up bug-finding and bug-fixing by shoving developers' noses into their products' flaws.

Kapor always had one particular customer in mind for dogfooding: himself. He needed a calendar that would allow both him and his assistant, Esther Sun, to make changes and keep their two copies synchronized. Chandler's vision of "read-write peer-to-peer sharing" was concocted, in part, to support this use case. They had tried using Apple's iCal; it turned out to be reasonably good at sharing the information from one calendar in a read-only way—essentially publishing it for wider consumption—but it simply wasn't flexible enough to give two people "read-write" privileges.

If Chandler was going to be dogfood-usable, it had to provide a calendar that Kapor and Sun could share. That meant making Chandler's broad promise of intuitive, versatile information-sharing real. But sharing very much remained what the Chandler team had labeled it a year before: a snake. The design team hadn't nailed the myriad open issues that plagued it, and the programmers still didn't know how they were going to build it.

There simply weren't many clear road signs—examples of successful systems to work from. Lotus Notes, developed by Ray Ozzie at Kapor's old company after Kapor left, was one, but it relied on a heavyweight server, and OSAF was committed to a server-free, peer-to-peer approach. After Lotus, Ozzie had gone on to build a new sharing and collaboration system called Groove (Kapor was an early investor). Groove was a true peer-to-peer system, and it had solved many of the thorniest problems of that design—including keeping users' data in sync and maintaining security (a critical matter for the military customers Groove increasingly served in the post-9/11 era). But Groove was Windows-only software, and OSAF was committed to allowing Mac and Linux users to share on an equal footing.

The further they delved into the problem of sharing, the more everyone at OSAF began to wonder whether they had imposed one too many conflicting requirements on themselves. Maybe there was a reason nobody had ever built software that seamlessly shared information across multiple computing platforms without relying on servers. Maybe it just wasn't practical. Maybe OSAF needed to rethink its plans.

■ ■ ■

That was how it looked to Lisa Dusseault. Dusseault had joined OSAF in March 2004 and began sharing the job of managing the developers with Katie Parlante. Within a few weeks they split the apps team, which was growing unwieldy, in two: The reconstituted Apps Group would continue to build Chandler's GUI, the front-end interface, while the new services team would focus on the layers of Chandler above the repository but below the GUI—functions like email transport and sharing mechanisms—that interacted with other software behind the scenes.

Dusseault was a bright, methodical, enthusiastic young Canadian with precise diction, sidelines in karate and knitting, and a passion for the abstruse realm of Internet standards. As a college student in the early nineties, she had interned at Microsoft and went to work as a program manager

there straight out of school in 1996. At that time, Microsoft was busy ripping out the guts of Windows to build Internet access and Web browser
features into the heart of its software. (The effort would later serve as a
central issue in the company's antitrust battle with the U.S. Department of
Justice.) Dusseault worked on several different server projects, including
early versions of the Microsoft Exchange Server that would later bedevil
Mitch Kapor's weekends.

Dusseault's arrival marked a number of firsts for OSAF. She was the
first full-fledged Microsoft veteran to join the project. She had left the software giant in 2000, disillusioned by a new boss ("he was ex-military and
operated on a need-to-know basis"), and had managed development at a
small start-up company for several years. She was also the first member of
the Chandler team with much firsthand experience of the Internet standards world, a loose bureaucracy that governs the approval of network
engineering standards—widely agreed upon, open, well-documented protocols that allow anyone to build compatible software. As part of her job at
Microsoft, Dusseault had started participating in the work of standards
committees and attending meetings of the Internet Engineering Task Force
(IETF), the parent organization of the standards committees.

At the IETF, Dusseault got involved in work on a new standard called
WebDAV (for Web-based Distributed Authoring and Versioning), an
extension to the basic protocol of the Web that, as its official Web site
explains, "allows users to collaboratively edit and manage files on remote
Web servers." By 2002 she was chairing the WebDAV working group and
writing a book on the subject.

To many programmers the realm of standards is a swamp of endless
draft revisions, arm wrestling via mailing list, and maddeningly slow
progress that makes garden-variety software development look like the
Indy 500. (Big software companies are notorious for implementing new
standards long before the drafts that codify them have won approval.) But
Dusseault took avidly to this world despite its quicksand pace.

"I think the reason I do all this is that I look a little bit longer than most people. I see the life cycles," she says. "The time frame most people are used to having is one year. And that's a good thing for software developers. Working on code that you don't envision being used for five years? It's so hard to know how to do it right. One year is probably even too long. So when they take that event horizon and go into standards work, and it takes the working group two years to go through the process of considering the draft, minimum, then another six months to approve it—and then you've probably not done the whole thing, you've just done the framework, now you have to do the next part of it. It just seems ridiculously overlong. You have to have an event horizon of five years. It'll be five years before something useful comes of this that you can really point to."

That didn't throw Dusseault. "In the standards world, I found a collection of techies who were thinking big. They weren't thinking about the release next summer or the bug due to be fixed next week. They were thinking about altering standards that had already existed for fifteen years, like email, or inventing new standards that they hoped would exist for twenty-five years."

To do that, Dusseault says, you have to think more like a city planner. "Talk about the email system: It's an ever changing collection of pieces of software and actual servers and people's accounts and everything, with remarkable stability. It's one of those almost natural systems that certainly didn't get designed by anybody. Pieces of it got designed and put in place and then morphed. So a big part of my approach is, whenever you're putting into place a new piece of a system, always look at whether you're going to regret it in five or ten years—whether you can extend it, replace it."

Soon after her arrival at OSAF, Dusseault gave her colleagues a lunchtime tutorial about WebDAV. She described it as a "stealth protocol": it's built into both Microsoft Windows (as Web Folders) and Apple's Macintosh OSX, and it is used inside many software packages that need to provide some form of remote sharing and editing of documents. But it's not well known. "People pick up WebDAV and use it because it looks right. It fits. You don't need to pull and push on it too much to make it work for what you're trying to do."

WebDAV works by extending HTTP—the protocol that Web servers and browsers use to talk to each other—adding new commands that allow users to edit files on a remote server. In specifying these extensions, WebDAV's creators had dealt with several of the problems that had caused Chandler's developers to tear their hair out as they tried to implement sharing. For instance, WebDAV had built-in mechanisms for conflict resolution— deciding what happens when two different users try to make simultaneous changes in the same document. Or you could avoid conflict in the first place through "locking," which allows a user editing a document to make sure that no one else has access to it until the work is done and the changes are saved.

If WebDAV could do it, why was it so hard for Chandler? Chandler's peer-to-peer approach meant there was no central server to be what developers call, with a kind of flip reverence, "the source of truth." WebDAV's server stored the document, knew what was happening to it, and could coordinate messages about its status to multiple users. Under a decentralized peer-to-peer approach, multiple copies of a document can proliferate with no master copy for users to rely on, no authority to turn to.

Life is harder without a "source of truth." For programmers as for other human beings, a canonical authority can be convenient. It rescues you from having to figure out how to adjudicate dilemmas on your own. After just a few weeks at OSAF, Dusseault became convinced that the peer-to-peer road to Chandler sharing was likely to prove a dead end. The project had little to show from its efforts to date anyway. "There wasn't any working peer-to-peer code," she later told me. "But it was like, we're doing peer-to-peer. We have to. We said we would. We decided to."

Programmers always bring their preexisting enthusiasms and expertise to a new problem. At worst this can lead to mismatches of the "when you have a hammer, everything looks like a nail" variety; at best it means that when you bring new people into a project, you get a free ride on their hobbyhorses. Whichever it was, from the moment Dusseault walked in the door at OSAF, it seemed a good bet that Chandler would end up using WebDAV somehow. And in fact, soon after joining the project, Dusseault began quietly lobbying for a new approach to Chandler's "sharing transport"—a WebDAV approach.

That would mean using a server, a big about-face for OSAF. It took Kapor some time to get used to the idea. There had always been a server somewhere in Chandler's future for the Westwood edition of the program, the version intended for use on university campuses. But Kapor knew that going with a server for the first edition of Chandler, the one they had been calling Canoga, was a real compromise and might look like a cave-in. Outsiders would be disappointed; OSAF developers might resist. But if Chandler was ever going to support sharing—and it had to—WebDAV had one overwhelming advantage: It already worked.

Kapor was rethinking the merits of peer-to-peer anyway. At a June meeting where he formally announced the adoption of a server-based sharing design to his staff, he explained, "I've had a significant change of point of view. There was a kind of frontier idealism that was well intentioned but not practical on my part. The issue is about empowering people. It's not about the infrastructure. Maybe we need a robust public infrastructure of servers to let people do what they want to do. My and OSAF's original position was, electricity is good, therefore everyone should have their own power plant! Unconsciously, I always imagined that user empowerment somehow meant a server-free or server-light environment. Now I think that's actually wrong."

A few weeks later he elaborated: "This actually turns out to be a deeply charged issue. So many of the people who are thought leaders in open source value freedom and initiative, and those values have been very tied up with this American frontier myth of self-sufficiency—going out on your own and just doing it. And while there's a lot to say for that, it turns out that the reality of open source and the Internet is much more collaborative than the narrow libertarian P2P ethic."

For at least some of the Chandler developers, peer-to-peer wasn't an ideology or a religion, but it was a major part of the vision they had signed up for, and some of the code they had already written was designed to support it. So selling the new approach to the development team was tricky. Andi Vajda, the repository developer, was especially resistant.

"We have to choose a path forward," Katie Parlante explained at a meeting in June. "We can't do all the neat stuff at once," Dusseault said.

But running a server meant asking users to hand their data over to somebody else, Vajda objected. "I don't use .Mac. I don't use Gmail. I don't trust a third party with my data. Anyway, to me it's not like sharing anymore. It's more like you're publishing this stuff."

Vajda had been building the Chandler repository with an eye to supporting direct repository-to-repository syncing and sharing. As he saw it, this was a basic requirement of the project. Kapor had always said that he wanted Chandler to make it easy for users to store one set of data on a work machine and one on a home machine, and keep the two copies in sync.

"Mitch has just taken away one reason to use Chandler as an end user if you don't do the peer-to-peer stuff," Vajda told me later that day. "So that's frustrating. And it's going to get worse because there will be more of these—more things where we say, 'Oh, this is too hard. We'd better give up on it.' Ultimately, they're right, because it's better to ship something that's not done but that's usable than wait ten years until you're completely done. The intent is good. But the vision is suffering, and that's frustrating. With this new server thing, Chandler just got bigger. People don't realize that. They think, Ah, it's easier. It's simpler than what repository sharing could do for them. . . . Maybe I'm wrong. I love to be wrong when I understand why. That's great! I'll learn something. But I'm a little worried, because the project just got bigger, not smaller."

Under the new WebDAV-based plan, when a Chandler user decided to share a collection of items, like a calendar, Chandler would copy that collection over to a WebDAV server. The "sharee" would log in to that server and have access to the calendar—to read it or change it. The server would make sure that all changes would then propagate back to the sharer. There were still lots of unanswered questions: How would I let you know I wanted to share with you? How would the server control access to different users' data? How would you go about terminating a share? But the biggest question was the one of "Chandlerness."

Chandler items weren't simply little bits of text. They were software objects with sets of attributes that defined what they were, what collections they belonged to, and how they related to one another. WebDAV had been designed to coordinate collaboration on Web pages and similar text documents. What would happen to the Chandler items when they got exported to the WebDAV server? Could they retain their Chandlerness? How much extra work would it take?

The only way to answer those questions was to begin to play around with the code. Meanwhile, Lisa Dusseault was already working on a new project, labeled CalDAV, that would extend WebDAV to make it smarter about handling calendars. If Chandler used CalDAV, and CalDAV emerged as a standard that other calendar software supported, Chandler users could end up sharing their calendars with people using other calendar programs. Eventually, Dusseault and Parlante could see creating a further set of WebDAV extensions—call it ChanDAV—that would allow WebDAV to handle all the dynamic, fluid data generated by Chandler.

For now, the task was simpler: Set up a WebDAV server for OSAF's developers to use, and write the code that would send Chandler items back and forth to the server. The rest of the vision, with CalDAV and ChanDAV and further DAVs yet unnamed, remained far down the road.

But with a realistic plan for sharing in place, there was one fewer snake hissing on the pavement.

■ ■ ■

It's a Monday afternoon late in June 2004. Chao Lam and Lisa Dusseault step into Mitch Kapor's sunlight-flooded office and sit down around a small round table near the door. Kapor joins them. Their agenda: planning for Canoga, the eventual 1.0 release of Chandler, which still seems like a distant peak to scale.

Recently, the managers and developers at OSAF have been experiencing another *Groundhog Day* repeat. They poked their heads up from their monitors, looked around, and realized that four months had passed since the release of Chandler 0.3, so there was no way they were going to release 0.4 in the three or four months allotted by their schedule. Now Kapor, Lam, and Dusseault are assessing what this latest slippage means for their long-term plans.

"Are we feature-driven or schedule-driven?" asks Dusseault.

"It has to be a little bit of both," Lam says.

"In the real world, it's always some of both," Kapor agrees. "It's pathological if it's all one or the other. Our plan is that the feature side is dominant. Of course we'll trim and adjust in iterative fashion. It's like a binary star system in which one is bigger than the other. They both influence each other's orbits.

"I'm being driven by dogfood at this point. I'd rather get something that is usable sooner. The argument against doing that is, it risks releasing something that's not all that interesting at a particular point in time. But that criticism is outweighed by 'half a loaf is better than none.' "

They begin reviewing the list of potential Canoga features that Lam has wrangled into Excel. It's full of major projects that the developers haven't even begun to tackle—things like importing and exporting data from Chandler and printing and "versioning" (keeping track of changes that users make so that they can undo them). As they consider each additional topic, the energy seems to sap out of the room.

"When I got here, I immediately identified sharing as an iceberg because it was an area I was really familiar with," Dusseault says. "There are other areas where I'm not familiar enough to know whether there's another iceberg under the water."

"Couldn't you do a best-case, no-icebergs schedule?" Kapor asks.

Even then, building a full set of features into Chandler is going to take time, Dusseault answers: there's only so much you can do to accelerate things, and if you hire too many new programmers, Brooks's Law will kick in.

"That's true," Kapor admits. "It's the old 'It takes nine months to make a baby, no matter what' principle."

"I think Canoga is one to two years from now," Dusseault says. "At least."

Kapor lets out a long, slow sigh. "What informs that?"

"The number of things that scare me. And things that I know aren't hard but are just a lot of slogging. I'm comparing it to how fuzzy the features felt on Exchange Server when it was two years before shipping, and then how many features they cut."

"And that's just Canoga," Kapor says. He blows air through his lips like a long, silent whistle, closes his eyes, and clasps his hands to his mouth. There's a long silence. He makes a halfhearted effort to challenge Dusseault's logic: "It's hard to compare what the rate of progress is going to be. We have new people just starting out now who are working on solving email and the WebDAV piece to start with. The character of the work changes after 0.4. . . ."

"But I also have Katie's team's plan for the next four months," replies Dusseault. "All the preferences dialogues and the printing dialogues—they take time. You just have to slog through them."

Kapor gives up, moves on. "The next question is, when do we have the usable dogfood release?"

"Dogfood starts really small with a couple of dedicated people."

"My hunch would be another six months of solid work if we really focused on dogfood. I'm sure it would involve a bunch of painful decisions."

"Another six months after 0.4?" Dusseault asks, with a look that says, respectfully, *I don't think so!*

"Yeah. What can we say by September for the CSG meeting [the university consortium] about what's dogfoodable by next May?"

Dusseault gazes out the window, out across Howard Street. "When people ask for numbers that far out, the traditional thing that engineers do is make them up."

In meetings like this one, I usually sit off to the side, unobtrusively taking notes. But now Kapor turns to me and jokes, "Scott is naming this chapter Realism Sets In."

I nod and smile, and think, *It's software time all over again.*

■ ▮ ▯

```
11:34    hazmat >    you know what would be really
                     helpful to outside developers
                     is a diagram of the chandler
                     components/packages and their
                     various states of stability
```

```
11:34    katie >     hazmat: sure, we could
                     probably arrange something
                     like that even for next
                     week . . .


11:36    hazmat >    that would be great, there is
                     so much changing and its
                     rather hard to know whats not
                     in flux


11:36    katie >     hazmat: dude, its all in flux!
```

—From the Chandler IRC chat channel, July 14, 2004

A handful of people outside OSAF showed up regularly on the IRC channel to follow the project and ask the core developers questions. On rare occasions the Chandler crew had even received patches—small bits of code correcting some problem—from outside developers. Chandler remained open source in the sense that its source code was public, as was the work taking place on it, but OSAF still didn't have many bazaar-style volunteer contributions propelling its work. "People seem to be extraordinarily patient with us," Ted Leung wrote on his blog, "and I hope that their patience will turn into active support and contribution once we get to a semi-usable state."

The inside developers kept at their spadework. David Surovell was working full-time on making fixes and improvements to the wxWidgets components that the apps team needed. He even made some headway toward fixing the infamous "flicker" bug that John Anderson had filed back in the days of the 0.1 release. It turned out to be a different issue on each of the three operating systems under which Chandler ran, having to do with how each system handled "double buffering" of the images on the screen—a way of preparing the visual contents of a window in a kind of off-screen holding pen so that they're ready to display the moment the user's clicks require it.

Meanwhile, Donn Denman was wiring up the detail view and getting stamping to work. Brian Kirsch, the email specialist, had integrated a Python-based open source framework called Twisted into Chandler. Twisted handled many different kinds of networking and could take care of much of the behind-the-scenes business of sending and receiving emails for Chandler. And John Anderson was busy "refactoring" the code for CPIA blocks to make it more compact and efficient.

"For instance," he said, showing the staff meeting one day, "this repository view was 121 lines, now it's 85."

"So the metric of how many lines of code you write is a tricky one," Kapor pointed out. "We've just done negative 36 lines of code."

Two new additions to the staff promised to move OSAF toward more structure and discipline in its development process. In July, Aparna Kadakia, a veteran of Netscape and other Silicon Valley companies, became OSAF's QA (quality assurance) manager and began to work on setting up a more thorough testing program for Chandler. And in August, Sheila Mooney—a programmer turned manager who had run projects for a medical software company and, before that, had worked on multimedia at Microsoft—joined Chao Lam on the product management side with a mandate to get serious about writing more formal and detailed specifications for different aspects of Chandler.

Still, there were big holes in the major features planned for the 0.4 release, and it was already August. Stuart Parmenter, who had taken on the job of getting rudimentary sharing via WebDAV working for 0.4, had a long way to go. John Anderson would make progress and then lose whole weeks to problems he referred to as "ratholes." And the clock was ticking; it was six months since the 0.3 release. The 0.4 schedule called for a feature freeze in early September, then a period of bug fixing and documentation, and a late October completion date. Kapor, Parlante, and Dusseault all knew that they had missed their dogfood target.

For several months they had been working to track tasks and schedules by individual developer for each two-week milestone release of the code. Was it helping to propel Chandler forward? Hard to say. But they knew it was taking a lot of their time.

■ ■ ■

"The 'not done' gap is pretty substantial," Kapor says at a review meeting on August 10 as the managers run down the long list of the latest milestone's targets, met and unmet. "The goal is to be better forecasters."

"Let me be the devil's advocate for a moment," Parlante says. "Always meeting plan is less important than having a high rate of progress and knowing where you're at. If we water down the plan so that we meet our goals more, that could make us less ambitious, and we'd get less done."

"People are bad at estimating the percent that's done," Dusseault suggests. "They're better at days remaining."

Kapor says carefully, "Let me underscore that the point of tracking all this is to be able to coordinate, not to praise or blame."

"More of the current process isn't going to help Stuart finish," says Dusseault. "It won't fix the not-doneness."

"One of the pitfalls you can fall into," Parlante says, "is the engineers say 'I'm done!' And then Aparna and the design team look at it and say, 'What are you talking about?' " In other words, the feature may "work," but there are tons of bugs, and the interface is nothing like the design, and a mountain of labor remains.

Ted Leung, who patches into these meetings over a conference room phone pod, speaks up: "The point of writing this stuff down is, we're trying to steer. You can't do that without some information. The question is, how much information do we need?"

"I'm trying to notch up the impatience level just a little bit," Kapor says. "Without being a jerk. There needs to be room to feel good about what was accomplished even as we accept that some things didn't get done. Otherwise, we'll just stick our heads in the oven and turn on the gas."

The custom in these meetings is to report team by team—apps, services, repository. After the troubled accounts of the other groups, Leung's report on the repository work—two small projects under way, both on schedule, nothing surprising—comes as a relief. Kapor takes note.

"So it is interesting how, earlier in the project, we spent endless amounts of time agonizing over the repository, involving everyone, sometimes bringing things to an agonizing halt for months at a time. Now we're trying to do all these other things that we've never done before and hitting roadblocks. But it's moved up the stack. I think it's helpful to remember that.

"For me as a software designer, it would be great if you could imagine a feature and it would just appear magically. But right now we're in hell. We have so many good design ideas that we'll be taking out and shooting—regularly.

"The good news is, the last time I tried something like this, which was at On Technology, I had shut it down before we got this far. This time I'm committed to launching."

■ ■ ■

Meanwhile, eight hundred miles to the north, a parallel software-time drama was playing out. Microsoft had been ballyhooing its new Longhorn operating system since the summer of 2003, but—big surprise!—the company began reining Longhorn in as 2004 progressed. By summer, Microsoft let it be known that the new version of Windows would not be available until late 2006 and would not contain many of the features it had touted so proudly the year before—including the fancy new WinFS file system that had sounded so similar to Chandler's siloless approach to data.

But while both Longhorn and Chandler slipped, a sea change was under way in the Web world: A new generation of software products was emerging with Web-browser-based interfaces, and, unlike their predecessors, geek enthusiasts gleefully declared, they *did not suck*. On April 1, Google, the phenomenally successful search engine, had unveiled a new Web-based email service called Gmail. It was so unexpected a move that the April Fool's launch date caused many observers to dismiss the announcement as a prank. But it was real. And while much of the stir around Gmail centered on the unprecedentedly large one-gigabyte size of the free mailboxes Google was handing out, the service's real innovation lay in its "dump everything in one place" approach to information management. Instead of sorting your mail into folders, you would just keep one big pile of mail; Gmail would intelligently sort the threads and help you find things with a great search tool, which is Google's forte.

Google had, in Chandleresque fashion, taken a crowbar to at least one set of information silos. It also provided a spare interface, which responded more quickly to a user's clicks than most Web-based programs by relying on some ingenious programming techniques. Gmail cut out the need for the browser to trade messages with the server for every single action you chose to perform. (That problem had figured prominently in Mitch Kapor's rationale for not building Chandler as a Web-based application.) It made the Web browser work a little more like the kind of "rich client" application Chandler was going to be. It was whittling down the difference between desktop applications and Web applications.

Gmail wasn't the only product to show off such wizardry, which later became known by the acronym AJAX (for asynchronous Javascript and XML); smaller outfits were performing similar cool tricks. Flickr, the product of a small company based in Vancouver, British Columbia, achieved the same kind of speed and delight in a new Web-based service that allowed people to share digital photos. And Basecamp, developed by a little Chicago design firm called 37 Signals, offered similar features in a Web-based tool for small-group project management.

Such products didn't instantly seize a mass market, but they won the embrace of enthusiastic early adopters—geeks and technology writers who raved about them in their blogs and helped spark the sense that a whole new generation of software was on the rise. Kapor had always taken the long view—a healthy approach for dealing with the technology industry's fickle fads—but he knew that Chandler wasn't growing in a vacuum and that its success depended on its promise of a markedly better approach to organizing information.

Several friendly observers of OSAF that I talked to expressed the fear more bluntly: If Chandler didn't hurry up, it risked being irrelevant.

On September 7, Katie Parlante, Lisa Dusseault, Chao Lam, and Sheila Mooney gather at the end of a long day of meetings for what they're calling a "kibble planning session."

"Kibble" is the informal name the Chandler team has adopted for their product's first dogfoodable release—not yet Canoga but a version of the software that they can use themselves every day. What exactly will be in it? How long is it going to take? And how will they get there?

In two weeks, Kapor and Lam are set to meet with OSAF's backers at CSG, the university consortium. They want a clear plan to present. Lam has emailed his colleagues an enormous spreadsheet that exhaustively compiles Chandler's entire set of features and then breaks them down into the scores of individual tasks that need to be completed to get them to work.

It is an eye-glazing document. The four sit around the big conference table and stare at the grids on their laptop screens.

"How much time will it take to review this stuff?" Dusseault asks. "Will it take the entire next two weeks?"

"Do you have an alternative?" Parlante replies.

Dusseault ponders the spreadsheet's rows and columns. "We need some way of understanding how big a release is, of putting features in similar size buckets. This format doesn't really allow us to compare things. The only real data point we have is how much we were able to achieve in past releases.

"One thing we could do," she suggests, "is put it all on stickies, where each note represents roughly even granularity." One Post-it note would represent one or two months of a single developer's time. They would lay the notes out on the wall organized by "dot release." Then they could see at a glance how much they had planned to accomplish in each round and whether they were being realistic.

"Stickies on a whiteboard?" Sheila Mooney says. "I've done that before. We did everything that way at my last job."

For a minute it seemed like they're going to schedule that exercise for yet another meeting—it's after 5:00 P.M. already. But instead Dusseault runs out and returns with a pile of Post-it pads. She clears the left third of the long whiteboard at the front of the room and draws four vertical columns on it, labeled 0.4, 0.5, 0.6, and 0.7. Then she draws two more horizontal lines, dividing the columns into top, middle, and bottom, corresponding to each of OSAF's three development groups—apps, services, and repository.

Dusseault, Parlante, and Mooney start marking up stickies and affixing them to the whiteboard. They confer about names, placement, and whether to divide some extra-difficult tasks into two stickies. "Is core versioning too big an item?" "Should search be one or two notes?"

The board fills up with clusters of yellow stickies inscribed with black markers. A smaller number of blue ones represent back-end tasks— programming jobs that don't involve any user interface work. A smattering of purple ones stand for work involving the project's infrastructure, like the source code control system and the build system.

Within fifteen minutes they have fifty notes up. By the half-hour mark it's pushing one hundred.

Mooney turns around with a grin: "Technology at work!"

As the managers work from Lam's list, they're able to cross off a number of features that had already been canceled or postponed or that simply aren't necessary to achieve dogfood—ideas like a "fisheye view" that Mimi Yin had proposed, which would concentrate the presentation of information on the most relevant items, or "item conversations," little instant-message-like annotations that users could append to shared items. Each feature they drop means at least one fewer sticky note or maybe more. But by the end they've still got 138 notes: 27 under 0.4; 45 under 0.5; 33 under 0.6; and 33 for 0.7 and beyond.

They pull chairs over from the conference table, sit in front of the sticky-laden board, and stare.

Mooney speaks up: "It's particularly sad that the 0.4 bucket"—the one representing work completed on the nearly finished current release—"has the least things in it." Meanwhile, the 0.5 bucket is overflowing.

"And we intended 0.5 to be shorter than 0.4," Dusseault says. "So it looks like our kibble plan will take at least three more 0.4s. And 0.4 has taken us eight months."

"I don't think they're all apples-to-apples comparisons," Parlante objects. "There was new people digestion. There was a pretty long planning phase."

"I still think two years for kibble is not unreasonable, based on what we're trying to do," Dusseault says.

Lam lifts his chin from his hands. "I think Mitch has the idea that it's sometime next year."

"I assume if we say it's two years to do all that, we're going to start cutting some things out of what we call kibble," Dusseault replies. "Don't forget, kibble is whatever we say it is."

"The biggest guiding principle is that at least we ourselves within OSAF can use it day to day," says Lam.

"Then some of that we don't need to do. Like PDA sync"— synchronization with mobile devices (Personal Digital Assistants).

"That depends if you're a PDA user or not."

"Two years is a minimum for this number of stickies."

"You're also assuming we don't get better at it."

"But you don't get increased productivity per programmer. Remember *The Mythical Man-Month*. Also remember the increased number of bugs to fix as we have more code to maintain."

Mooney says, "We also have how many new people we're hiring?"

"I don't want anyone to think that we can make 0.5 come in any faster by hiring," Dusseault says. "That would be a mistake. I was imagining that the people you're adding would be covering the increased maintenance. I find it hard to account for either of those, so I'm kind of hoping that they cancel each other out."

"What if we were hiring at twice the rate?" Lam asks.

"Over the first year it would slow us down," Dusseault says. "Two years, even. Three years, it starts to pay off."

"Then scratch that." Lam smiles. "Okay. Well, this was useful. At least we have a number."

Lam and Dusseault were reenacting an archetypal dialogue: the tug-of-war between product manager and development manager. It has occurred in countless meeting rooms at countless software projects. It always goes like this:

PRODUCT MANAGER: Too long! Can't we
                       move faster?

DEVELOPMENT MANAGER: This is how long it takes.

PRODUCT MANAGER: Can't we do better?

DEVELOPMENT MANAGER: If we're really good, we can know how long it will take, but we can't change it.

Two years to kibble. Less, if they cut features and lower their sights.

They agree that Lam will break the news to Kapor. No one expects him to greet it with joy.

As they file out, Mooney writes, in big capital letters at the top of the whiteboard: DO NOT ERASE.

■ ■ ■

On Thursday, as the OSAF staff assembles for the next "all-hands" meeting, the swarm of sticky notes still covers one end of the whiteboard. With a thicket of yellow notes along the top, a thinner band of blue ones below, and an occasional purple one scattered about, it looks a bit like the screen of an old Space Invaders video game, minus the gun at the bottom to pick off the rows of targets. It can't be missed. It demands explanation.

Kapor waves his hand toward the board as the meeting begins. "This represents the most sophisticated, state-of-the-art, long-term engineering management tool." Laughs ripple through the room. "It also represents an accomplishment. It's a useful exercise to start blocking out the major chunks of work to get us to kibble. The good news is that the process now exists. Work is being done on it. But it certainly appears that we need to be rigorous about eliminating nonessential features in order to get where we want to be in a reasonable amount of time."

Next, Chao Lam reviews a list of product features that may become casualties of the new push for "kibble sooner": Contact management (your address book). Instant messaging. Mimi Yin's iTunes-like browser. The month view in the calendar. Treating email attachments as real items in Chandler.

"These are like my puppies," Kapor says. "This feels like putting some of them on the shelf."

The room is silent. It's hard to tell whether the programmers are relieved that, well, realism has set in—or disappointed at how much has been axed.

"It would be fair to say that some of the innovation we really wanted to do will slip out of the first release. We bit off more than we could chew. The product is going to be more vanilla than we had originally hoped and more similar to its predecessors. On the other hand, there's still more than a critical mass of important new features: the dashboard view, getting away from silos, and sharing. Our claim to be doing new interesting stuff will still stand up. It's just going to be challenging to get things narrowed down enough to get a product out sooner rather than later."

In a presentation to OSAF's small board of directors later that month, Kapor reviewed a set of explanations for why Chandler was so late: They had underestimated the cost of the project's big ambitions. It had proved harder than expected to build an engineering organization. Developing "rich client" software was hard everywhere—"It's not just us." And on the personal side: "I knew less than I thought. Or I was rusty."

No one could accuse Kapor of ducking the least pleasant test of leadership—coming clean about failures and failings. But he had a way of acknowledging setbacks without letting them derail or demoralize him. He had taken the bad news from the stickies meeting in stride. He wouldn't sugarcoat it for his staff; they were too smart for crude spin anyway. But he framed it with a kind of reality-tempered optimism—offering a glimpse of the qualities it must have taken two decades before to build little Lotus into a successful giant.

■ ■ ■

The development managers and Kapor debated their options over the next week. Kapor challenged the assumption that each subsequent release would take as long as 0.4's eight months and began to lay out alternative scenarios. One idea was to drop the idea of a Canoga 1.0 release entirely. Instead, the developers would work single-mindedly toward the dogfood-able kibble release, which could be refined into a respectable Chandler 1.0 once it had been field-tested by OSAF and brave souls beyond. Then they would march on to the university-focused Westwood release.

The other big idea was to postpone serious work on implementing email and concentrate on building a usable calendar. Chao Lam first proposed this in a memo titled Outrageous Questions, and it raised hackles. It felt as if, in Dusseault's words, it might be "gutting the vision too far." Parlante pointed out that much of the work needed to produce a dogfoodable Chandler lay in simply making the application stable enough to use, and that had to be done regardless of what order they built the chunks of the PIM functionality (calendar, email, tasks, and notes).

But Kapor argued that postponing building email into Chandler could help them deliver working software sooner, and that was the right priority. So they drew a new line across the stickies board to separate all the tasks required to produce a functioning calendar from everything else, and began a process that came to be known as "stickie refactoring"—unpeeling and repasting notes.

How much could they accomplish in a year or so to produce a working kibble release by the end of 2005?

Dusseault pointed to the above-the-line areas on the whiteboard, the calendar-only features under the 0.5, 0.6, and 0.7 headings: "The stuff that's here, here, and here we can do in a year. And also some of this." *This* was one bucket of the rest of the Chandler features below the calendar-only line.

At the next all-hands meeting, on September 16, Kapor laid out the new program: No Canoga. Dogfood or bust. Calendar features first. "The good news is, there's a growing optimism that we can ship something meaningful by the end of 2005. It's still a work in progress. It would be a mistake to say we have a working plan. But that's our hypothesis."

The news was hardly a shock to anyone in the room, but it was sobering nonetheless. There was still so far to go. But for the first time, at least, they could see they had a plan grounded in reality, rooted in estimates that bore some relationship to the record of how long it actually took them to implement a feature in code. They were no longer planning "top down," based on hopes and pronouncements; they were planning "bottom up," based on experience and evidence.

Through all the sail-trimming in this period, Kapor had been finding solace in a phenomenon entirely outside OSAF: the rising tide of success that the Mozilla Project had found with its Firefox browser. Firefox, a streamlined, modernized version of the old Mozilla software that had evolved from the ashes of Netscape, had a 1.0 release scheduled for later that fall. But even in its prerelease version it was logging a million downloads a week and winning rave reviews. It was free open source software. And it had emerged out of a project whose trajectory eerily resembled Chandler's. Mozilla had begun with great fanfare and enthusiasm among the geek elite and then dropped into a black hole for years in which it lost the allegiance of some key developers and was widely written off. Over and over again, it seemed, Mozilla had missed the boat; with Firefox, its ship finally came in.

Mitchell Baker, the former Netscape lawyer whose efforts as an administrator and manager helped keep Mozilla alive through its dark years, had continued to serve as a kind of community adviser to OSAF. As the meeting of Chandler developers tried to digest the discouraging news from Kapor about their schedule, she spoke up: "In the early phases of the Mozilla Project, there were discussions similar to this. But it was pretty dysfunctional. We lost a lot of time, some of it to the technology but a lot of it to the inability to get management—which in those days was Netscape management—connected with the key engineering forces."

"How did you get on the right path?" Pieter Hartsook, OSAF's marketing guy, asked.

Baker talked about the Mozilla developers' decision to rewrite from scratch the browser's "layout engine," the code that draws Web pages on screen. The choice cost the project years, and outsiders often criticized it. Still, she said, "just making that layout decision was key—even if it might have been the wrong one. And then figuring out an engineering process that worked. And then the third thing was, just getting far enough along to have something that basically works."

Kapor nodded. "Does this sound familiar?"

"Like every software project I've ever worked on," John Anderson said with a fatalistic smile.

■ ■ ■

OSAF remained a congenial workplace. The lunches were good, the hours were humane, and the atmosphere was unfailingly nice. But delay and retrenchment took some toll on the staff. Brian Skinner had gone on leave at the start of the summer, frustrated by the slow pace and discouraged that he had never been able to carve out a piece of the project to call his own. He decided not to return. As the fall progressed, Stuart Parmenter announced he was going to leave, too—to join an open source calendaring project that Oracle was sponsoring. Around the same time Chao Lam decided to scale back his involvement; he would keep his hand in Chandler but wanted to explore starting up his own company again. Sheila Mooney stepped into his role as project manager. Soon after, Ducky Sherwood decided to go back to school for a computer science degree.

Motivating programmers and their managers through the ups and downs of a project's long life cycle is a mysterious art. For all the turnover at OSAF, there still seemed to be enough magnetism around Chandler to hold talented people. But it wasn't the relatively mundane work on the calendar that kept them charged up.

Chandler had always led a double life: In one vision it was a software application for users to manage information, and in the other it was an "application framework" or "platform," a foundation on which developers could keep adding cool new stuff. Since Kapor and the OSAF team had gradually and grudgingly come to accept that they couldn't do everything at once, they had chosen to put the application first. "Working software sooner!" They knew from experience that this was probably the smart choice. But they still dreamed about building the bigger platform—the open-ended, world-changing tool that, with the right additions of code, could organize and connect all sorts of disparate data.

Lisa Dusseault was fired up about hooking Chandler to CalDAV, making it interoperable with a wider universe of calendar software. Donn Denman was about to start a six-month paternity leave, but when he got back, he was eager to begin building a scripting function into Chandler that would eventually allow nonprogrammers to automate it. Andi Vajda, who had already spun off the work he had done to integrate Python and the Lucene search engine into a new open source project under the OSAF umbrella named PyLucene, dreamed of turning the Chandler repository into a more independent piece of software that other projects might be able to use.

At the same meeting where Denman showed off his working stamping code, Morgen Sagen presented another demo to the staff. He had wanted to start a photo blog—a Web diary of photographs—for his family and wasn't happy with the available software, so he "decided to roll my own with Python," he explained. He needed a database to store the photos. And Web-serving code. "And I realized, Wait a second! What project do I know that already does all this?" (The Twisted networking code that was now incorporated in Chandler came with its own Web server.)

So as his colleagues watched, rapt, Sagen showed his new photo blog—pages filled with pictures of his young children—built on top of Chandler, with the photos stored and organized inside the Chandler application but published on the Web for anyone to see.

It was nothing extraordinary in itself. And it was still crude and unfinished. But for this group of people it served as a tantalizing preview of a future they had been working toward—one that still felt painfully distant.

Somebody asked, "How long did it take you to do this?"

"About two hours," Sagen replied.

CHAPTER 9

METHODS

R emember the quality triangle—time, money, and features or quality? By the fall of 2004, two years after the first public announcement of Chandler, OSAF wasn't faring especially well on any of its three sides. Fast? Hardly. Cheap? Several million dollars and, so far, not much to show. Good? That remained to be seen.

When Chandler 0.4 was released on October 26, 2004, about one thousand people downloaded it the first week compared with the fifteen thousand who had rushed to check out Chandler 0.1 in the first twenty-four hours. The latest work in progress was beginning to look a little more like a real software application when you started it up, and it had some new functions built into it that you could test if you were willing to dig. It had a partially working detail view for the first time, and you could stamp items to change them from one type to another.

But for basic everyday tasks—sending or receiving email, storing calendar information, and organizing notes—Chandler still didn't work well enough for anyone to try to use it. It wasn't even dogfoodable by its own programmers. It took forever to start up, and it crashed a lot.

No wonder the download numbers were down. For now at least, the world had moved on. Many outsiders who had cheered the project's launch simply gave up on it, and some former participants felt it had lost its way. "It looks like a complete train wreck to me at this point," Andy Hertzfeld said, ruefully shaking his head, when I talked to him soon after 0.4's unveiling.

But for the first time there was a sense of calm at OSAF surrounding the release. The weeks leading up to it were the opposite of frantic. The managers and QA staff (of one) methodically reviewed the list of open bugs in Bugzilla and "triaged" them: This one's getting fixed this week; that one's trivial. This one we'll punt to the next release; that one is a bona fide "blocker" that must be fixed. The apps team—which was responsible for the aspect of Chandler, its user interface, that had the most bugs to fix— methodically whittled down the list. The design team began planning 0.5 well before 0.4's completion, and the developers who weren't hip deep in bugs began working on their 0.5 projects.

OSAF may not yet have had a product to deliver to the world. But it had something else: the beginnings of a process, a working methodology that might carry it toward that goal.

Uh, wait a minute, you may well say. They had been working for two years *without* a process? That's not exactly true. OSAF had plenty of processes, maybe too many, and the processes sometimes seemed as fluid and open to revision as Chandler's code. But something changed during the eight months of 0.4; the project finally cut down on improvisation and developed some structure—organizational machinery that could be counted on to roll forward without having to be rebuilt every few months or needing its engine restarted for each new release.

From the earliest days of Chandler, Kapor had been adamant about trying to draw up plans and schedules that were honest and realistic, and yet when it came to meeting schedules, the project had a poor record. It was reliably averaging one release every six months, yet its plans continued to assume releases could be finished in three or four. Part of the problem, perhaps, was that in software development, as in all things, plans get dodgier the farther into the future one looks. As Lisa Dusseault put it, "You have to be aware that whatever your current plan is, if it's looking a year out, it probably really sucks."

In encountering this discouraging reality, Chandler was no different from the great majority of software projects. It's rare for a group of software developers to work together on a series of projects over time; in this they are less like sports teams or military units or musical ensembles and more like the forces of pros who assemble to make a movie and then disperse and recombine for the next film. So, while individual programmers and managers may carry with them a wealth of experience and knowledge of techniques that served them well in the past, each time they begin a new project with a new team, they are likely to end up pressing the reset button and having to devise a working process from first principles.

Since a successful process is tough to figure out—and even after companies or teams come up with one, they usually lack the institutional memory to reuse it—the software profession has invented a whole host of "methodologies." Think of them as freeze-dried processes, recipes for software success. Some programmers swear by them—sometimes one at a time, sometimes promiscuously. Many others look down their noses at the very idea that their unique talents should submit to any procedural yoke.

From the earliest days of the industry, advocates of one or another methodology have promised that it alone holds the key to bringing software projects in on time, on budget, and with fewer bugs. The problem with this one-size-fits-all rhetoric is that no one approach could possibly work across the wide range of software projects. What makes sense for a product destined for use by millions of consumers (the old shrink-wrapped software market) probably doesn't make sense for software written to be used internally at one company. A program destined for use by an entire government agency makes different demands on its developers than a specialty product for, say, audio postproduction. Software for use on personal computers is a different beast from software for use in cell phones or embedded electronics—the systems that run inside cars or security systems. Prescriptions for best practices don't offer refills. Silver bullets won't reload.

■ ▮ ▯

As I followed Chandler's fitful progress and watched the project's machinery sputter and cough, I kept circling back to the reactions I'd had to my own experiences with software time: *It can't always be like this.* Somebody must have figured this stuff out. Somewhere there is a map with a path out of the software-time tar pit.

I began to hunt for answers, picking up a lantern and wandering the corridors of software history and analysis, knocking on doors and asking, "Is there anyone who can honestly say he knows a better way to make software?"

My search started with the 1960s, the era in which the creators of software first turned their systematic minds to their field and themselves. In 1968, Edsger Dijkstra published a brief piece in *Communications of the ACM* with the puckish title "Go To Statement Considered Harmful." In many of the dominant programming languages of the era, like Fortran and Basic, the computer moved sequentially through a list of numbered instruction statements, and the programmer could jump from one point in the sequence to another using a statement called GOTO. Like a film editor making a jump cut, GOTO simply dropped one story line that a program was developing and picked up another. It handed off control from one point in a program to another unconditionally, taking nothing else into account—neither the values of different variables nor the state of the program and its data.

"The go to statement as it stands is just too primitive," Dijkstra wrote. "It is too much an invitation to make a mess of one's program."

In the sixties and seventies, many programmers accepted that invitation and made messes—confused tangles of software that came to be known as "spaghetti code." Taking their cue from Dijsktra and other theorists who had begun to envision a more modular style of programming, advocates of a methodology known as "structured programming" proposed a set of practices to banish the plague of procedural pasta. There were several competing schools of structured programming; in Dijsktra's, the central idea was to compose a program as a collection of subunits, each of which had only a single point of entry and a single point of exit. The program, Dijkstra wrote, could then be envisioned not as a pile of noodles but as a "necklace strung from individual pearls," whose simplicity and clarity not only reduced the likelihood of error but increased the possibility that programmers could understand their own handiwork.

Structured programming offered its recommendations from a defensive crouch, trying to protect fallible programmers from their own flaws. "As a slow-witted human being I have a very small head, and I had better learn to live with it and to respect my limitations and give them full credit," Dijkstra wrote. On winning the Turing Award, computing's Nobel, he titled his 1972 lecture "The Humble Programmer," and his writing is full of swells of pity for the "poor programmer" and lamentations over the human animal's "inability to do much."

The specific injunctions of structured programming would quickly be codified in the design and syntax of new generations of programming languages. Its broader imperative—that "each program layer is to be understood all by itself"—has driven one innovation after another in subsequent decades as programmers have devised new ways to isolate, "modularize," and "encapsulate" the moving parts of software from one another (with the dream of reusable Lego components as its ultimate grail). But like each of those innovations to come, structured programming didn't help perfect the software of its day; instead, it laid a solid foundation for more ambitious software to come, and those programs would find new ways to fail that structured programming couldn't imagine.

It turned out that improving how you organize code was a cakewalk compared with improving how you organize people and their work. As the industry's poor record held its depressing course through the sixties and seventies, the proponents of new software methodologies turned their attention to the human side of the enterprise. In particular, as they grappled with the tendency for large-scale projects to run aground, they began to focus on improving the process of planning. And, over time, they split into two camps with opposite recommendations. Like open source projects whose contributors have a falling out, they forked.

One school looked at planning and said, "Good plans are so valuable and so hard to make, you need to plan harder, plan longer, plan in more detail, and plan again! Plan until you can't plan any further—and then plan some more." The other school said, essentially, "Planning is so impossible and the results are so useless that you should just give up. Go with the flow! Write your code, listen to feedback, work with your customers, and change it as you go along. That's the only plan you can count on." I'm exaggerating, plainly, but the divide is real, and most software developers and managers fall by natural inclination or bitter experience on one side of it or the other.

Watts Humphrey, the guru of disciplined project management, lays out the "Must. Plan. More!" argument: "Unless developers plan and track their personal work, that work will be unpredictable. Furthermore, if the cost and schedule of the developers' personal work is unpredictable, the cost and schedule of their teams' work will also be unpredictable. And, of course, when a project team's work is unpredictable, the entire project is unpredictable. In short, as long as individual developers do not plan and track their personal work, their projects will be uncontrollable and unmanageable."

On the other hand, here is Peter Drucker, the father of contemporary management studies: "Most discussions of the knowledge worker's task start with the advice to plan one's work. This sounds eminently plausible. The only thing wrong with it is that it rarely works. The plans always remain on paper, always remain good intentions. They seldom turn into achievement."

Drucker published those words in 1966. As it happened, that was just about the time that a young Watts Humphrey was taking over the reins of software management at IBM.

■ ■ ■

IBM was in the middle of trying to deliver OS/360, an operating system for its new System/360 generation of mainframe computing—the fiasco that inspired Frederick Brooks's seminal writing in *The Mythical Man-Month*. Before accepting the job of running the OS/360 team, Brooks already had one foot out IBM's door; he was leaving to found a new computer science department at the University of North Carolina. As Humphrey tells it, although Brooks had ably led the 360 programming team through the concept and design phase, he departed before the ideas were implemented, and after he left, the project got into "serious trouble."

Serious trouble: Slippage. Delay. Revised schedules that slipped again. And more delay. IBM eventually delivered the System/360 hardware without the full new operating system. That took an entire extra year.

When Humphrey took over the IBM software organization in January 1966, he was stunned by its state, he wrote in a brief memoir.

> My first need was to find out where the work stood. . . . In each laboratory I visited, I asked management for their plans and schedules. No one had anything other than informal notes or memos. When I asked for the managers' views on the best way to manage software, they all said

they would make a plan before they started development. When I asked why they did not do this, their answer was that they did not have time. This was clearly nonsense! The right way to do the job was undoubtedly the fastest and cheapest way. It was obvious that these managers were not managing but reacting. They were under enormous pressure and had so much to do they could only do those things that were clearly required for shipping code. Everything but the immediate crisis was deferred. I concluded that the situation was really my fault, not theirs. As long as I permitted them to announce and ship products without plans, they would continue to do so.

Humphrey also found that IBM's marketing division was announcing products even when the software teams had no plans or resources to deliver them. So he went to his boss. "I told him that since all the delivery schedules were not worth anything anyway, I intended to cancel them. Next, I would instruct all the software managers to produce plans for every project. From then on, we would not announce, deliver, or fund programming projects until I first had a documented and signed-off development plan on my desk. . . . It took the laboratories about sixty days to produce their first plans. This group, who had never before made a delivery schedule, did not miss a date for the next two and a half years."

Humphrey's success at enforcing schedule discipline at IBM stood on two principles: Plans were mandatory. And plans had to be realistic. They had to be "bottom-up," derived from the experience and knowledge of the programmers who would commit to meeting them, rather than "top-down," imposed by executive fiat or marketing wish.

In the 1980s, Humphrey retired from IBM and wondered what he would do next. As he later told Business Week, "My daughter persuaded me to go to a seminar on commitment. At the seminar there was talk about making an 'outrageous commitment,' like stamping out world hunger. So I made an outrageous commitment that when I left IBM, I was going to change the way software was developed in the world. It certainly beats sitting on the beach."

Humphrey joined forces with the Software Engineering Institute (SEI) at Carnegie Mellon University, an outfit the Pentagon had founded in 1984 to try to improve the quality of the increasingly vast software systems it was buying with taxpayers' money. At SEI, Humphrey and his colleagues created the Capability Maturity Model (CMM) as a kind of yardstick for judging the quality of software development organizations.

The CMM provides a five-step ladder for programming teams to climb. You could spend the rest of your life reading all the documents that describe CMM principles, but the informal description that Humphrey provided in 1997 offers a simple sketch:

> An organization at Level 1 is basically not doing much of anything. At Level 2, they're doing some planning, tracking, configuration management, they make some noises about quality assurance, that kind of stuff. A Level 3 organization begins to define processes—how they work, how they get things done, trainable things. At Level 4 they're using measurements. They have a framework for actually tracking and managing what they do, something statistically trackable. Level 5 organizations have a continuously improving process.

The CMM served best as a tool for the Defense Department to measure the organizational prowess of contractors. It found less traction in the commercial software world and the burgeoning personal computer software industry. To this day if you mention the CMM in most small software companies, you will receive blank stares. Among those programmers and managers who have paid some attention to the CMM, the most common criticism is that it's impersonal and bureaucratic and imposes heavy, innovation-stifling burdens. In a 1994 article titled "The Immaturity of CMM," software quality expert James Bach set out to debunk the CMM:

> The CMM reveres process, but ignores people. . . . Innovation per se does not appear in the CMM at all, and it is only suggested by level 5. . . . Preoccupied with predictability, the CMM is profoundly ignorant of the dynamics of innovation. . . . Because the CMM is distrustful of personal contributions, ignorant of the conditions needed to nurture non-linear ideas, and content to bury them beneath a constraining superstructure, achieving level 2 on the CMM scale may very well stamp out the only flame that lit the company to begin with.

At least in part as a response to such complaints, Humphrey and the SEI devised two sequels to the CMM. The original methodology aimed chiefly to help very large organizations improve software schedules and quality; the follow-ups—the Team Software Process (TSP) and Personal Software Process (PSP)—targeted these problems on a smaller scale. TSP and PSP criticize "autocratic management styles" and encourage individual developers and small teams to seize control of their own destiny by taking responsibility for planning and quality control, sharing information, and "dynamically rebalancing" their workloads as needs change. One Humphrey presentation offered these bluntly persuasive bullet points:

- We all work for organizations.
- These organizations require plans.
- Unless you are independently wealthy, you must work to a schedule.

▸ If you don't make your own schedules, somebody
else will.
▸ Then that person will control your work.

The CMM, TSP, and PSP all drew inspiration from the ideas of manufacturing quality expert W. Edwards Deming, who argued that quality should not be an afterthought but ought to be built into every stage of a production process. In software this means the CMM is hostile to the "code and fix" tradition, where programmers produce bug-filled products, testers find bugs, and then programmers go back and fix them. "In all of modern technology," Humphrey says, "software is the only field that does not consider quality until products are put into test. It's not the engineers' fault, however, since this is how they've been trained and how they are managed."

Just as Deming's ideas were embraced by the postwar Japanese auto industry long before they won attention in his native United States, so the CMM has won its most avid following abroad—particularly in India's burgeoning software industry, where companies find that achieving a particular CMM level reassures overseas customers that they can be trusted with outsourced work. But the CMM and its related methodologies have yet to make a major dent in the world of business software or desktop computing in the United States.

A 2000 book titled *The Software Conspiracy: Why Software Companies Put Out Faulty Products, How They Can Hurt You, and What You Can Do About It* argues that this is a scandal. Since these methods are proven to produce better results, why doesn't every software team use them and every customer demand them? Author Mark Minasi spreads the blame liberally to self-indulgent programmers, shortsighted companies, and a public too willing to tolerate shoddiness and bugs.

Minasi and other CMM advocates maintain that the industry knows the answer to producing high-quality software on time and under budget. It's just too fat, dumb, and lazy to embrace it. They could be right. On the other hand, their arguments assume that the biggest challenge facing the makers of software is to reduce the number of bugs in their products. In the work of a typical experienced programmer, according to Humphrey, you'll find one bug for every seven to ten lines of code; the CMM and related methods, SEI studies show, reduce these numbers dramatically.

In the language of the CMM, TSP, and PSP, borrowed from the terminology of factory management, bugs are called "defects," and defects are "injected" by workers—as if the product begins in some Platonic state of perfection and is then corrupted by mad-eyed workers shooting bug-filled syringes into its bloodstream. These methods emphasize squashing bugs as soon as they are encountered because, according to a principle known as Boehm's Law, the later in the development process you fix a bug, the more it costs to do so.

"Fix bugs first" is a fine principle, and most programmers will agree that it is fundamentally the way to go—in principle. But whatever intentions programmers start with, their code is always born with bugs because the process of writing it is, to use one of the most popular words in the software development vocabulary, "iterative." You write something once so you can get started; you go over it again and again—you iterate—to improve it. If you insisted that it be perfect right out of the gate, you'd never get past the first line. In practice, "fix bugs first" works fine until the people who are waiting for the finished product grow impatient. Then the principle falls by the wayside in a scramble to deliver working, if imperfect, code.

Beyond that, there are bugs and there are bugs. If you scan the bug list that any active software project maintains, you'll find everything from showstoppers to trivialities. Bugs vary enormously in how much they matter to the user, how hard they are to find, and how hard they are to fix. And every bug list represents a vast exercise in apples-and-oranges comparison.

The annals of software bugs feature a small but still alarming number of dramatic disasters. Five minutes after its launch in July 1962, the Mariner 1 spacecraft veered off course, and flight control blew it up to prevent it from crashing in inhabited areas. The problem? A missing hyphen in its guidance control program. In June 1996, the European Space Agency's $500 million unmanned Ariane 5 rocket exploded forty seconds after liftoff because of a bug in the software that controlled its guidance system. (It tried to convert a 64-bit variable to a 16-bit variable, but the number was too high, a buffer overflowed, and the system froze.) From 1985 to 1987 a radiation therapy machine named the Therac-25 delivered massive X-ray overdoses to a half-dozen patients because of software flaws. During the 1991 Gulf War, a battery of American Patriot missiles failed to fire against incoming Scud missiles; the enemy missile hit a U.S. barracks, leaving twenty-eight dead. Investigations found that the software's calculations had a defect that compounded over time, and after one hundred hours of continuous use, the Patriot's figures were so far off, it simply didn't fire.

These and similar stories are few enough in number not to spark a media frenzy or a political ruckus, but they serve as cautionary reminders of our dependence on the quality of programmers' work. And for every example of a fatal bug there are a thousand instances of bugs that cause property loss or financial damage. Software problems have been blamed for blackouts, airline delays and airport closings, bank account foul-ups, auto recalls, and virtually every other kind of snafu and inconvenience you can imagine.

But these disaster-causing flaws are not the "one per every ten lines of code" bugs; if they were, no software system could possibly work. While counting bugs is plainly a valuable exercise—it's better than *not* counting them—it tends to throw critical system failures and picayune flaws into the same bucket. It also ends up encouraging programmers to perfect existing products rather than build new things. Which is fine—unless building something new is what you care about.

■ ■ ■

In the 1980s and '90s, the U.S. military embraced CMM and the rest of the Software Engineering Institute's alphabet soup as valuable tools for improving the software it needed. But at the turn of the millennium, as the Pentagon's software planners pondered massive and futuristic new weapons programs and faced the pressures of post-9/11 military commitments, they began to worry that even with the help of these methodologies, the size of the software the new systems required was growing faster than programmers could keep up.

In a 2004 talk at the Pentagon's annual software development conference in Salt Lake City, Jon Ogg, director of engineering and technical management at the Air Force Materiel Command, laid out what he called the "software divergence dilemma" that the military faces today. In the past fifty years, the amount of code in a typical military system has increased a hundredfold. For instance, a 1960s-era jet fighter might have had fifty thousand lines of code, whereas the new Joint Strike Fighter will demand five million. Meanwhile, in that same span of time, the average productivity of programmers has only doubled. That means we've seen the "person-months of effort now required to develop a capability" increase by a factor of fifty, Ogg said.

At the same event, Barry Boehm—professor of software engineering at the University of Southern California, a pioneer in the field of cost estimation, and the father of Boehm's Law—put it more succinctly: "The amount of software the Department of Defense needs is growing exponentially, and you'll never get it done in a finite amount of time."

Confronted with this dilemma, the military's software suppliers, encouraged by their most important customer, the Pentagon, have begun to experiment with less traditional approaches. They're not alone. During the same years that the CMM and related methodologies were taking root among these defense contractors, an alternative tradition was evolving in the world outside, emerging directly from the experience of programmers who knew that their process was broken and yearned to find a better way.

One taproot of this movement lay in the work of the "Patterns community," the software developers inspired by the work of the architectural philosopher Christopher Alexander. In the late eighties and early nineties, the rise of object-oriented programming, with its dauntingly complex abstractions, was pushing many programmers to the limits of their comprehension. The software patterns movement, whose leaders included wiki inventor Ward Cunningham and a programmer named Kent Beck, imagined a new kind of lifeline for them, a less prescriptive approach to software methodology. Instead of laying down fixed principles of best practices, they recorded their experiences in brief narratives. "Faced with this kind of problem," they would say, "we've found this pattern of programming to be useful." The patterns movement approached software development as a craft; it held, as veteran computing columnist Brian Hayes wrote, that "programmers are like carpenters or stonemasons—stewards of a body of knowledge gained by experience and passed along by tradition and apprenticeship."

Patterns advocates like Cunningham and Beck brought hands-on practicality to a field that desperately needed it. For instance, to tame the complexity of object-oriented coding, Cunningham and Beck proposed that programmers design a new program by laying out index cards—one per software object—on a table. "These have the advantages that they are cheap, portable, readily available, and familiar," they wrote. "We were surprised at the value of physically moving the cards around. When learners pick up an object, they seem to more readily identify with it and are prepared to deal with the remainder of the design from its perspective. It is the value of this physical interaction that has led us to resist a computerization of the cards."

The patterns movement gave individual programmers a new way of thinking about solving problems and of distilling their learning for their colleagues' benefit. But it did not go far toward helping them figure out how to coordinate their work on larger projects.

For decades the organization of the typical project followed the "waterfall model." The waterfall approach—the label first surfaced in 1970—divided a project into an orderly sequence of discrete phases, like requirements definition, design, implementation, integration, testing, and deployment. One phase would finish before the next began. This all seemed logical on paper, but in practice it almost invariably led to delay, confusion, and disaster. Everything took forever, and nothing worked right. Programmers would either sit idle, waiting for the requirements, or give up and start design work and coding before they got the requirements. "Big design up front" led to big delays, and "big-bang integration"—writing major chunks of code separately and then putting them all together near the end of the project—caused system collapse. By the time the finished product arrived, so much time had passed that the problems the program aimed to solve no longer mattered, and new problems clamored for solutions.

The waterfall model gradually acquired the bad reputation it deserved. In the mid-eighties, Boehm defined an alternative known as the "spiral model," which broke development down to "iterations" of six months to two years—mini-waterfalls dedicated to producing working code faster and allowing feedback from use of the resulting partially completed product to guide the next iteration. The spiral became standard in the realm of large-scale government-contracted software development, where a typical project's "acquisition cycle" might span as long as a decade.

For the accelerating world of commercial software, however, it was still too slow. In the nineties, the software industry's methodology devotees adopted the banner of Rapid Application Development (RAD), which promised to speed up the delivery of finished software through quick prototyping, more aggressive iteration cycles, and reliance on new tools that let the computer itself handle some of programming's more mundane tasks. RAD helped software companies work more nimbly. But soon after it began to take hold, along came the Web, which ratcheted up the industry's hunger for speed yet again, to the manic pace of Internet time.

The fast new approaches of the nineties, which favored adapting to a project's constant change over trying to predict and control its outcome, came to be called "lightweight methodologies." That distinguished them from the plodding "heavyweights" of the world of the CMM, but it didn't help sell them to business leaders; it sounded wimpy. In 2001, a group of seventeen leaders in the field, including Cunningham and Beck, gathered at a Utah ski resort to try to find common ground among their related but diverse approaches. Brian Marick, an expert in software testing methodologies who participated in the event, wrote: "Part of the purpose of the workshop . . . was to find a replacement for the term 'lightweight processes.' People felt that saying 'I'm a lightweight!' somehow didn't get the right message across."

The meeting found a more virile name for the movement—Agile Software Development—and produced a manifesto that reads in its entirety:

> We are uncovering better ways of developing software by
> doing it and helping others do it.
>> Through this work we have come to value:
>
> ▸ **Individuals and interactions** over processes and
>   tools
> ▸ **Working software** over comprehensive docu-
>   mentation
> ▸ **Customer collaboration** over contract negotiation
> ▸ **Responding to change** over following a plan
>
>> That is, while there is value in the items on the right,
> we value the items on the left more.

In a field whose theoretical documents are prone to acronym overkill and bureaucratic bloat, the Agile Manifesto is remarkable for its brevity and simplicity. But "agile development" was more of an umbrella for shared values than a roadmap of specific processes. A variety of related but distinct agile methodologies have flourished since the manifesto's publication. One, Scrum, divides projects into thirty-day "sprints" and emphasizes daily meetings to keep work on track. The most popular species of agile methodology by far, though, is Extreme Programming (or XP, not to be confused with the Windows XP operating system).

Kent Beck developed the ideas behind Extreme Programming in the 1990s while consulting for Chrysler on ways to rescue a payroll automation project that had run off the rails. The intentionally provocative name sounds like something that involves the purchase of a lot of high-priced sporting goods, and in fact Beck has drawn comparisons between the sort of dedication and courage "extreme sports" require in athletes and the demands Extreme Programming places on coders. But Extreme Programming's label mostly refers to the way it adopts a set of widely accepted methods and then pushes them to their limits.

"We were taking all these practices that the best programmers already did and turning the dials all the way up to 10" is how Ron Jeffries, one of XP's creators, once put it. Is testing important? Then have developers write their tests *before* they write their code. Is it good for the development team to talk with the customer? Then keep the customer on hand to answer the developers' questions. Instead of periodic code reviews, have developers work in pairs all the time so that every line of code has more than one set of eyes on it from the moment it is written. Above all, accept that the customer's requirements, and thus the software's goals, are going to keep changing, and organize your project to "embrace change."

XP introduced a new and sometimes strange vocabulary to the software development world. It mandated breaking projects down into "stories." Each story represents a feature request that the customer lays out for the developers in a narrative that explains what the program should do. The programmers go off and code the "story." If it does what the customer wants, they're done and ready for the next story. Under XP's radical incrementalism, there is no "big design up front." Coding starts almost immediately. This appeals to many programmers, whose itch to start cranking out code is often thwarted by lengthy design processes. The license XP grants to forget about writing detailed specifications and documentation of code is also popular.

But programmers often find some other XP tenets difficult to follow, especially the rallying cry of You Aren't Gonna Need It, or YAGNI. YAGNI takes the just-in-time manufacturing method, in which factories don't collect the parts to build a product until a customer places an order, and applies it to the world of knowledge goods. The principle advises, "Always implement things when you actually need them, never when you just foresee that you need them." In other words: Your meeting-scheduling program may someday need to support users in different time zones, but if your company has only one office right now, that can probably wait. This is practical but painful counsel for the habitual axe sharpeners of the programming tribe.

Conventional wisdom holds that agile development and XP are best for small teams of experienced coders. Though XP has found enthusiastic converts in pockets all over the software industry and even on a handful of teams writing code for the Pentagon, it's a demanding discipline if you try to obey all of its mandates. That tempts some programmers to cherry-pick only those rules they find congenial. Critics have argued that XP often serves as an excuse for lazy coders to ignore the discipline demanded by specifications, documentation, planning, and all the other onerous burdens of "heavyweight" process. XP can quickly decay into a means to, in the phrase of one open source expert, "Deliver Crap Quickly."

Once, while interviewing a job candidate at Salon, I apologetically described the state of disarray of our technical operation at the time.

"Ahhh," said the interviewee with a diplomatic smile. "Then you're using agile methodology!"

WILLARD: They told me that you had gone totally insane and that your methods were unsound.

KURTZ: Are my methods unsound?

WILLARD: I don't see . . . any method at all, sir.

—*Apocalypse Now*

Any developer who has been around the block will admit that the cavalcade of methodologies over three decades of software history has left the field richer and given programmers useful new tools and ways of thinking about their work. But finding a developer or team that actually subscribes to a particular methodology isn't easy. A 2004 study by two Pennsylvania software-engineering professors quizzed two hundred software team leaders in a cross-section of industries about their development practices. The study's authors wrote of "the shock and disappointment we felt at finding that the most dominant practice was none at all—a practice (if it can be called such) reported by a full third of the survey participants."

Shock and disappointment! Many software developers are still rebels, crying, "Don't fence me in." But many others, even while resisting the labels and the hype, quietly embrace and promote many of the techniques that different methodologies have recommended through the years. "This stuff is too hard to do without *some* systematic practices," they'll say.

Joel Spolsky is in this group. A former Microsoft project manager, he worked on Excel, the Microsoft Office spreadsheet that seized first place from Mitch Kapor's Lotus 1-2-3 in the early nineties and has held it ever since. When Spolsky left Microsoft to found his own small software company, he began a Web site called Joel on Software, where he publishes entertainingly tart essays full of pragmatic advice. Here is a taste of one:

> If you have even the slightest bit of common sense, you should ask: "Where's the data? If I'm going to switch to Intense Programming I want to see proof that the extra money spent on dog kennels and bird cages is going to pay for itself in increased programmer self-esteem. Show me hard data!"
>
> And, of course, we have none. . . .
>
> You can't honestly compare the productivity of two software teams unless they are trying to build exactly the same thing under exactly the same circumstances with the exact same human individuals, who have been some-how cloned so they don't learn anything the first time through the experiment. . . .
>
> We don't have any data. You can give us anecdotes left and right about how methodology X worked or didn't work, but you can't prove that when it worked it wasn't just because of one really, really good programmer on the team, and you can't prove that when it failed it wasn't just because the company was in the process of going bank-rupt and everybody was too demoralized to do anything at all, Aeron chairs notwithstanding.

Spolsky is highly skeptical of what he calls "Big-M methodologies." In another essay he writes: "Beware of Methodologies. They are a great way to bring everyone up to a dismal but passable level of performance, but at the same time, they are aggravating to more talented people who chafe at the restrictions that are placed on them. It's pretty obvious to me that a talented chef is not going to be happy making burgers at McDonald's, precisely because of McDonald's rules."

When I interviewed Spolsky several years after he had written that, he hadn't lost any of his disdain. "The real goal of a methodology," he said, "is to sell books, not to actually solve anybody's problem." According to Spolsky, "The key problem with the methodologies is that, implemented by smart people, the kind of people who invent methodologies, they work. Implemented by shlubs who will not do anything more than following instructions they are given, they won't work. Anyway, the majority of developers don't read books about software development, they don't read Web sites about software development, they don't even read Slashdot. So they're never going to get this, no matter how much we keep writing about it.

"Even today, as I look around, even though it's all on the Web, and there's a million books you can buy about how to manage, there's a million things you can read about how to make your software project work well, you still see the same old characters in Silicon Valley, [who] should know better by now, because it's their third company, making the same mistakes, doing the same things wrong, making the same silly assumptions."

Spolsky does have his own sorta-kinda methodology: He calls it the Joel Test and offers it as a sort of quick-and-dirty set of principles, based on his own experience and "collective wisdom," to judge whether a development organization has its act together—"CMM without the headaches." The Joel Test asks the following dozen questions:

> Do you use source control?
>
> Can you make a build in one step?
>
> Do you make daily builds?
>
> Do you have a bug database?
>
> Do you fix bugs before writing new code?
>
> Do you have an up-to-date schedule?
>
> Do you have a spec?
>
> Do programmers have quiet working conditions?
>
> Do you use the best tools money can buy?

Do you have testers?

Do new candidates write code during their interview?

Do you do hallway usability testing?

"A score of 12 is perfect," Spolsky wrote, "11 is tolerable, but 10 or lower and you've got serious problems. The truth is that most software organizations are running with a score of 2 or 3, and they need serious help, because companies like Microsoft run at 12 full-time."

Microsoft is, of course, the most successful software company in history, routinely throwing off billions of dollars in profits. Whatever it's doing is worth some attention. On the other hand, Microsoft is no less prone to late schedules and overlooked bugs than its commercial competitors or the legions of in-house developers who build corporate software. If Microsoft aces the Joel Test, why does it still have problems? "I think Microsoft is now well in the advanced throes of getting everything 100 percent right in terms of the discipline and just no longer producing products people want to buy with any level of consistency," Spolsky says. "The process is overtuned. . . . Now it's gotten to the point where it takes sixteen months to get out any release of any Microsoft product. They really can't steer that ship very fast. They spent something like a year on SP2 [Windows XP Service Pack 2]. It's a good thing they did that, but it's almost all cleanup, maintenance, catch-up, for security reasons.

"It's what the army calls fatigue. Fatigue is everything in the army that you do to keep your equipment in good working condition: polishing your shoes, brushing your teeth, making sure that you're ready and that all your bullets are clean and there's no sand in your gun. It's all called fatigue, and it takes about two hours a day for an infantry guy. And it's everything but the actual thing you're trying to do. Microsoft has now got to the point where it's like 80 percent, 90 percent fatigue."

Spolsky's foot soldier analogy comes from firsthand experience: He grew up in Israel and served in the military there. But the comparison comes up a lot. Software developers are frequently portrayed as foot sol-diers on death marches in pitched corporate battles. They take on missions and struggle to accomplish them under rapidly shifting circumstances and fog-of-war-style visibility. And they frequently develop that combination of fatalism and perseverance that has marked battlefield veterans throughout history.

At the Pentagon software conference I attended, I listened to a Personal Software Process trainer named David Cook explain the purpose of methodological discipline in terms that were one part pep talk and one part existential sermon.

"What are the odds your software is going to change?" Cook asked in a jovial blare. "Oh, 100 percent. It's going to be around for ten years. You're going to have to modify it more times than you can imagine. You're gonna have to change it for different hardware and software environments, change languages three times, and go through four personnel turnovers. What you have to do is fight as hard as you can to keep quality up so that ten years from now you have something left.

"It's like, when you grab the software in your hands, it's sand. And as you stand here doing the best you can, sand starts leaking between your fin-gers. Ten years from now there's a few grains of sand left, and that's all you have to work with. Your job is to keep your hands as tight as possible."

■ ▮ ▯

Despite the odds—despite complexity and delay and unpredictable change—a lot of software somehow does get written and delivered and, finally, used. Occasionally, it's even good. Rarely, it actually does something new and valuable. And in a handful of cases, it achieves all of that *on schedule*.

Very often, in those rare cases, success is a by-product of iron-willed restraint—a choice firmly made and vociferously reasserted at every challenge to limit a project's scope. Where you find software success stories, you invariably find people who are good at saying no. Like an artist who deliberately limits his palette to one color, a poet who chooses to write a sonnet instead of free verse, or a manufacturer who chooses to serve one small product niche, the successful programmer thrives because of, not in spite of, constraints. Sometimes those constraints are the product of circumstance—small budgets, tight deadlines, limited goals. Sometimes they are self-imposed by experienced developers or managers who have learned the hard way to shun open-ended—in software talk, "unbounded"— projects. Either way, the perspective is less "small is beautiful" than "big is dangerous."

A tiny Chicago-based company named 37 Signals is the poster child for this limit-embracing approach. 37 Signals began as a Web design consulting firm and branched out into software development at first simply to meet its own needs. It built some internal tools to manage projects. Then it needed a way to communicate with clients, so it opened up parts of its system to them. Before they knew it, as company founder and president Jason Fried explains it, they had the makings of a Web-based application. With four months more work, they turned their software into a service called Basecamp. Launched in February 2004, it quickly became a leader in the parade of new, rich, Web-based applications like Google's Gmail and Flickr.

Basecamp was the first in a series of noteworthy minimalist products that this company with only a handful of employees rolled out in little over a year. Next came Ta-da List, for saving and sharing to-do (and any other kind of) lists. A few months after that came Backpack, which allows users to save and share notes and files. Each product was solid, easy to use, and carefully designed, and each typically offered a small number of new twists.

Backpack, for instance, had some neat email capabilities. You could send email reminders to yourself, as lots of other services and programs allowed, but you could also send email to each of your Backpack Web pages from another computer or a mobile device like a cell phone, and the text of the email would appear on the page.

I started using Backpack to store random research notes for this book, and when I bumped into Fried at a technology conference in the fall of 2004, I asked him how 37 Signals managed to crank out so much useful software in such a short time. He is a passionate proselytizer for his approach—his company offers a workshop, "The Making of Basecamp," that distills its principles into a PowerPoint presentation—and obliged me with a forty-five-minute overview of his methodology right there in the hotel lobby.

For one thing, 37 Signals had only one developer, so it avoided the whole Brooks's Law quagmire—just as Mitch Kapor had in his original work on Lotus 1-2-3, which was mostly written by a single programmer, Jonathan Sachs. Coordination among developers simply wasn't an issue. It didn't even seem to matter that the lone programmer, a young whiz named David Heinemeyer Hansson, lived in Denmark. At most companies, Fried said, this geographical separation would be viewed as a terrible problem, but the time difference actually meant that they had only a few hours' overlap to talk things over, so they would use it efficiently and then the developer could code in peace without interruption.

Constraints, according to 37 Signals, are your friends. "Constraints are key to building a great product," Fried said. "They're what makes creativity happen. If someone said you have all the money in the world to build what-ever you want, it would probably never be released. Give me just a month!"

Another key to 37 Signals' ability to produce good software fast lay in its focus on Web-based applications. Everything worked through the Web browser, so the programs would work on any kind of computer under any operating system that could run a browser. Incremental changes could be rolled in easily on the server that ran the service, without users having to download and install new upgrades. Hansson was also keen on working with Ruby, a dynamic object-oriented programming language. Similar to Python but less widely known, it sped up and simplified Hansson's work. Finally, the 37 Signals approach shunned the ritual of writing up specs; instead, the work begins with detailed mock-ups of the Web pages that users will see. These page designs become the specs. Fried said it's not uncommon for his team to spend hours debating each word and button and box displayed on them.

37 Signals set out to create some small programs, not to build an ambitious new platform or application framework. But in the course of building Basecamp, Hansson had written some useful and innovative code that streamlined and simplified the basic chores that all Web applications had to perform in the course of storing and retrieving data. After Basecamp's launch, he and 37 Signals decided to take that work and release it as an open source platform called Ruby on Rails. Rails, as it came to be called, made writing Web applications easier, in part by limiting the programmer's options. "Flexibility is overrated. Constraints are liberating," Hansson says. And Rails was ready-made for the AJAX-style interface enhancements that were making those Web-based programs credible competition to their desktop equivalents.

Just as 37 Signals had extracted the Rails framework from the Basecamp code, it extracted a design philosophy from the Basecamp experience, encoded in a handy series of aphorisms: "Less software." "Say no by default." "Find the right people." "Don't build a half-assed product, build half a product." These are buzz phrases designed for quick consumption via presentation slides, but together they constitute a coherent approach to software development—call it pragmatic minimalism. It may not satisfy the yen to change the world that motivates so many programmers. You could criticize it as an indication of chastened ambitions. It doesn't seem to apply very well to projects that simply have no choice but to be big. As programmers like to say, it "doesn't scale."

But based on the 37 Signals track record to date, it has one great recommendation: In its own little way it does seem to work. A similar approach has found unprecedented success—even, you could say with some credibility, world-changing success—at a much larger and more celebrated software company. Google, which in a handful of years had grown into a billion-dollar behemoth and begun to challenge Microsoft, followed a software development philosophy that sounded remarkably like Jason Fried's: Small teams assembled ad hoc for each new project, worked on tight deadlines, and rolled out narrowly focused Web-based products, which they then improved incrementally based on user feedback and field experience. Google also told its programmers to devote one-fifth of their work time to personal projects. The fruits of these "20 percent time" labors might turn into cool new products—or not. Don't worry, the company reassured its people. Go forth and scratch your itches!

Google was gaining a reputation for having built a sort of engineers' paradise where algorithms ruled the roost and coders called the shots. Those lucky enough to be employed at the Googleplex—including Andy Hertzfeld and Python's Guido van Rossum, both of whom joined Google in 2005—found a working environment that, for a spell, had escaped the stasis of software time. Google had its share of half-baked products, but no one would dispute the value of its successes—from its original search engine to its keyword-based advertising business and its popular free email service.

Pragmatic minimalism had served Google well. But now it was a phenomenally successful public company facing new pressures to keep up the pace of growth and find new sources of revenue. To many it looked like Google was single-handedly inflating a new Silicon Valley bubble. If it could stick to its methodology under such circumstances, figuring out how to grow bigger without getting slower and dumber, it would prove unique in the history of the software industry.

I can't say that my quest to find better ways of making software was very successful. While I don't think the methodology peddlers are snake oil salespeople—they each report hard-earned lessons that software makers can learn from—they are inevitably backward-looking. They provide solutions for problems that beset their industry today but can't offer much help when the next wave of problems crashes down, which is why each new decade seems to bring a new bundle of methodologies.

But there is one set of circumstances under which methodologies really are The Answer. This is the scenario presented by a business thinker named Nicholas Carr in a notorious May 2003 article in the *Harvard Business Review* titled "IT Doesn't Matter." Carr infuriated legions of Silicon Valley visionaries and technology executives by suggesting that their products— the entire corpus of information technology, or IT—had become irrelevant. Like Francis Fukuyama, the Hegelian philosopher who famously declared "the end of history" when the Berlin wall fell and the Soviet Union imploded, Carr argued, essentially, that software history is *over*, done. We know what software is, what it does, and how to deploy it in the business world, so there is nothing left but to dot the i's and bring on the heavyweight methodologies to perfect it.

Drawing comparisons from previous generations of "disruptive technologies" like railroads and electricity, Carr argued that computers and software had at first offered farsighted early adopters an opportunity to seize comparative advantage. But now, he said, they have become commodities. They are important—you can't neglect them, or you'll fall behind the competition—but they won't "enable individual companies to distinguish themselves in a meaningful way from their competitors." In the grand game of corporate strategy and business competition, they simply don't matter. They've become boring infrastructure—mere plumbing.

The inflammatory title of Carr's article raised predictable hackles, but his argument made eminent sense as long as you viewed IT as a fixed set of functions and capabilities. IT "doesn't matter" as long as you know in advance that the things you expect software to accomplish for a business—accounting, workflow automation, inventory management, you name it—are not going to change and that your competition is using software to accomplish the same set of things.

But of all the capital goods in which businesses invest large sums, software is uniquely mutable. The gigantic software packages for "customer relationship management" and "enterprise resource planning" that occupy the lives of the CTOs and CIOs of big corporations may be cumbersome and expensive. But they are still—just as much as the little spreadsheets that Dan Bricklin and Mitch Kapor introduced to the world a quarter century ago—made of "thought-stuff" (to recall Frederick Brooks's term). And so every piece of software that gets used gets changed as people decide they want to adapt it for some new purpose. No sooner do you introduce a new system for employee surveys than the lightbulb goes off over somebody's head: Hey, why can't we turn the same system into a tool to schedule our company parties? All it will take is a few tweaks! And while they're at it, maybe the development team will build in a nifty way for me to catalog my music collection. Then we could vote on the music at the parties!

Have we already discovered and deployed all the usefulness—and, yes, all the frivolousness, too—that software offers? To believe so seems not only arrogant but actively ignorant of recent computing history. By the mid-1980s, for instance, business analysts were writing off the entire personal computer industry as mature and over the hill, just as the new wave of Macintosh and Windows-style graphic interface systems was building. In the early nineties, Microsoft was beginning to dominate the PC software industry, and the conventional wisdom was that the entire business was consolidating—yet the Internet was about to kick it in new directions. In 2000 and 2001, with the collapse of the Internet bubble, many observers were ready to declare the entire Net-based software industry dead and seal the coffin, but right around the corner Google was growing and helping to gestate a whole new wave of online businesses.

Over and over, those who have bet that there can be nothing new under the technology industry sun have lost their shirts. To believe that we already know all the possible uses for software is to assume that the programs we already possess satisfy all our needs and that people are going to stop seeking something better.

Irate critics of software flaws like *The Software Conspiracy*'s Mark Minasi and skeptical analysts of the software business like Nicholas Carr share these end-of-history blinders. If you believe that we already know everything we want from software, then it's natural to believe that with enough hard work and planning, we can perfect it—and that's where we should place our energies. Don't even think about new features and novel ideas; focus everyone's energies on whittling down every product's bug list until we can say, for the first time in history, that most software is in great shape. Microsoft has actually tried this. After a series of embarrassing virus outbreaks, for a time in 2003 it stopped most work on new products and committed much of its programming workforce to fixing Windows security holes. With the resulting XP Service Pack 2, Windows security improved, but that didn't stop people from inventing new viruses to plague it.

This stop-the-world mentality occasionally makes sense, particularly when flawed software that is already widely used cries out for fixing. But mostly, *it* is what's irrelevant: It misses the whole point of why we are still dissatisfied with the software we depend on. Users may be annoyed by bugs, and software developers may be disappointed by their inability to perfect their work, and managers may be frustrated by the unreliability of their plans. But in the end, none of that matters as much as the simple fact that software does not work the way we think, and until it does, it is not worth trying to perfect.

▮ ▮ ▮

Over lunch one day at OSAF, John Anderson confessed that he sometimes thinks software would be easier to write and would come out better if we could wave a wand and turn the clock back a decade with regard to the size of a computer's memory.

"With today's tools and processes, but with yesterday's memory limitations, we could really do a much better job," he said.

"But then you're just saying, let's set an upper limit on the ambition of software," I objected. "Let's have software that does less."

"Well, yes," he answered, "it might do less, but it would have fewer flaws."

▮ ▮ ▮

As long as we keep asking software to accomplish new things and solve new problems, fixing today's bugs still won't protect us from tomorrow's crashes. And although software methodologies are very helpful when you want to repeat a success, you can reproduce an outcome only if you have one to start with. If you are exploring unknown territory, best practices might help you move a little more quickly, but they're not going to point you to your destination.

Software development is often compared to the construction industry, but the analogy breaks down in one respect. Even after we have learned how to build structures to solve particular problems—sheltering a family, say, or providing an appropriate space for the treatment of sick people—we still need to keep building more of those structures. However much I may like your house, I can't have it myself. You're already living there, and if I want the same thing, I still have to build another one. Software doesn't work that way; once we know how to write a program to balance a checkbook or display a Web page, each additional copy of the program costs nothing to produce, and the price you pay, assuming you pay at all, is to reward the people who devised it, not to construct your copy. Therefore, once somebody has written a program that does what you need it to do, it's always cheaper to buy that software than build something from scratch.

Time and time again, however, we find that what we need or want is just different enough from what's available that we have no choice but to write new code. At Salon, for instance, in the late nineties we realized we needed to automate the way we published articles. The most popular software for content management at the time was a big package called Vignette, which had a six-figure price tag. A salesman from the company visited our offices and described its hundreds of useful features. Unfortunately, we had no need for most of them. But there was a handful of things we really did need, and Vignette couldn't do them. So we hired some programmers to write our own system instead.

Since every writer about software sooner or later ends up offering a law under his own name, the time has come for me, with all due humility, to present Rosenberg's Law: *Software is easy to make, except when you want it to do something new.* And then, of course, there is a corollary: *The only software that's worth making is software that does something new.*

■ ▮ ▯

The argument that "software is just plumbing" evokes an almost instinctive "Yes, but" reaction. Plumbing fulfills a limited and well-defined set of needs; we set the requirements a long time ago (hot water, drinking water, no leaks, clog-free toilets), and even when innovators try to change them—electronic toilet seats, anyone?—the changes often fail to take hold.

So far, in its brief but lengthening history, software is nothing like that.

Critics of the software state-of-the-art often complain that we are still in the "tinker with the engine" phase of this technology. In the early days of automobiles, the odds were good that your vehicle would break down before reaching its destination, so you had to be an amateur mechanic if you expected to get anywhere. Similarly, today's computer users must often be amateur programmers, learning to think more like the machine so they can fix the raft of problems they encounter when they try to get work done. Eventually, cars got more reliable, the controls got standardized, and you didn't have to be an obsessive hobbyist to do your driving. We are still waiting, but one of these days, this line of reasoning goes, something similar will happen with software.

It's a fair argument. Maybe someday we will get there, and software really will become just like your plumbing or your car—something we can manufacture according to routines, use without thinking, and expect never to change. But if the past is any guide, it's far more likely that we will keep expecting software to do new and different things. The car of one decade will turn into a flying car the next—then a submarine! And we will welcome, even demand, those metamorphoses.

CHAPTER 10

ENGINEERS AND ARTISTS

F rom a panel of experts assembled at a conference on software engineering, here are some comments:

"We undoubtedly produce software by backward techniques."

"Particularly alarming is the seemingly unavoidable fallibility of large software, since a malfunction in an advanced hardware-software system can be a matter of life and death."

"We build systems like the Wright brothers built airplanes—build the whole thing, push it off the cliff, let it crash, and start over again."

"Production of large software has become a scare item for management. By reputation it is often an unprofitable morass, costly and unending."

"The problems of scale would not be so frightening if we could at least place limits beforehand on the effort and cost required to complete a software task. . . . There is no theory which enables us to calculate limits on the size, performance, or complexity of software. There is, in many instances, no way even to specify in a logically tight way what the software product is supposed to do or how it is supposed to do it."

"I think it is probably impossible to specify a system completely free of ambiguities, certainly so if we use a natural language, such as English."

"Some people opine that any software system that cannot be completed by some four or five people within a year can never be completed."

These grim perspectives sound familiar, but I left out one detail: This wasn't a recent conference. Every word I have quoted, and thousands more like them, can be found in the proceedings of a 1968 summit on software engineering—the first event of its kind, organized by NATO as its member nations stared into the eyes of what they had just begun to call "the Software Crisis."

After IBM's OS/360 disaster and other problematic large-scale software projects, the NATO partners, still jousting in a global Cold War that had extended its competition into outer space, decided that software development had become an urgent international problem. They gathered dozens of their best minds in Garmisch, Germany, a secluded Bavarian ski resort, and asked them for some insight on the subjects of software reliability, quality control, cost control, and scheduling—the topics we refer to today under the catchall phrase "software engineering."

In the decades since that conference, these two words, *software* and *engineering*, have become fused at the hip, like *government* and *bureaucracy* or *fast* and *food*. At vast corporations and tiny start-ups alike, any employee who can throw together a quick script to automate something is given the title of software engineer. Not everyone is happy about this. In 2003, some programmers in Texas who referred to themselves as software engineers on their business cards received cease-and-desist letters from the state's Board of Professional Engineers, which jealously protects use of the label. But despite such bureaucratic rearguard actions, the term has become universal, only rarely raising questions about whether the parallel between building structures and building code makes sense.

That was not a foregone conclusion in 1968. When the Garmisch organizers named their event a conference on software engineering, they "fully accepted that the term . . . expressed a need rather than a reality," according to Brian Randell, a British software expert who organized the conference's report. "The phrase 'software engineering' was deliberately chosen as being provocative, in implying the need for software manufacture to be based on the types of theoretical foundations and practical disciplines, that are traditional in the established branches of engineering." The conference attendees intended to map out how they might bring the chaotic field of software under the scientific sway of engineering; it was an aspiration, not an accomplishment.

Yet something apparently happened in the interval between the 1968 conference, which left participants enthusiastic and excited, and its successor in Rome the following year. The second gathering aimed to focus on specific techniques of software engineering. "Unlike the first conference . . . in Rome there was already a slight tendency to talk as if the subject [software engineering] already existed," Randell wrote in a later memoir of the events. The term *software engineering* had almost overnight evolved from a "provocation" to a fait accompli.

In the original conference report, Randell wrote that the Rome event "bore little resemblance to its predecessor. The sense of urgency in the face of common problems was not so apparent as at Garmisch. Instead, a lack of communication between different sections of the participants became, in the editors' opinions at least, a dominant feature. Eventually the seriousness of this communications gap, and the realization that it was but a reflection of the situation in the real world, caused the gap itself to become a major topic of discussion."

The divisions that arose in Rome, along with a debate about the need for an international software engineering institute, led one participant, an IBM programmer named Tom Simpson, to write a satire titled "Masterpiece Engineering." Simpson imagined a conference in the year 1500 trying to "scientificize" the production of art masterpieces, attempting to specify the criteria for creation of the Mona Lisa, and establishing an institute to promote the more efficient production of great paintings.

> They set about equipping the masterpiece workers with some more efficient tools to help them create master-pieces. They invented power-driven chisels, automatic paint tube squeezers and so on. . . . Production was still not reaching satisfactory levels. . . . Two weeks at the Institute were spent in counting the number of brush strokes per day produced by one group of painters, and this criterion was then promptly applied in assessing the value to the enterprise of the rest. If a painter failed to turn in his twenty brush strokes per day he was clearly underproductive. Regrettably none of these advances in knowledge seemed to have any real impact on master-piece production and so, at length, the group decided that the basic difficulty was clearly a management problem. One of the brighter students (by the name of L. da Vinci) was instantly promoted to manager of the project, putting him in charge of procuring paints, canvases and brushes for the rest of the organisation.

Simpson's spoof of the Rome conference's work had teeth: "In a few hundred years," he wrote, "somebody may unearth our tape recordings on this spot and find us equally ridiculous." It was sufficiently painful to the conference organizers that they pressured Randell and the other authors of the report to remove it from the proceedings. (Years later Randell posted it on the Web.) "Masterpiece Engineering" could not have been more prescient in laying out the central fault line that has marked all subsequent discussions of the Software Crisis and its solutions—between those who see making software as a scientific process, susceptible to constant improvement, perhaps even perfectible, and those who see it as a primarily creative endeavor, one that might be tweaked toward efficiency but never made to run like clockwork.

The two 130-page reports of the NATO software engineering conferences foreshadow virtually all the subjects, ideas, and controversies that have occupied the software field through four subsequent decades. The participants recorded their frustration with the lumbering pace and uncertain results of large-scale software development and advocated many of the remedies that have flowed in and out of fashion in the years since: Small teams. "Feedback from users early in the design process." "Do something small, useful, now." "Use the criterion: It should be easy to explain."

George Santayana's dictum that "those who cannot remember the past are condemned to repeat it" applies here. It's tempting to recommend that these NATO reports be required reading for all programmers and their managers. But, as Joel Spolsky says, most programmers don't read much about their own discipline. That leaves them trapped in infinite loops of self-ignorance.

■ ▪ ▫

We tend to think of the labors of art and those of science as distinctly separate endeavors, but in fact they form more of a spectrum. In varying proportions, science and art share custody of most works of creation and insight; "masterpiece engineer" Leonardo da Vinci is perhaps the most famous reminder of that. But the software field has long been cursed with an overabundance of absolutists who insist that the solution to all its problems lies in ridding it of all human imprecision and sequestering it in the realm of pure science. At their most extreme, proponents of this view demand that we subject our programs to mathematical proofs; only programs that have been "proven correct" can be considered finished. I have not found many practicing programmers who see much value in this approach. (And as we will see in the next chapter, it may well be theoretically impossible anyway.)

Engineering is, of course, all about bridging the gulf between art and science. Engineering is often defined as the application of scientific principles to serve human needs. But it also brings creativity to bear on those scientific principles, dragging them out of pristine abstraction into the compromised universe of our frustrations and wants. The word derives (via a detour through medieval French) from the same Latin root that gave us *ingenious* and refers to the ability to make things skillfully. Today we use the term *engineering* in diverse ways, and we yoke it to other words to denote an array of subdisciplines: mechanical engineering, civil engineering, electrical engineering, structural engineering, chemical engineering, and so on. Software engineering seems no more than a latecomer to this list. Why is the concept so problematic?

As we have seen, software sometimes feels intractable because it is invisible. Things that we can't see are difficult to imagine, and highly complex things that we can't see are very difficult to communicate. But invisibility isn't the only issue. We can't see electricity or magnetism or gravity, but we can reliably predict their behavior for most practical purposes. Yet simply getting a piece of software to behave consistently so that you can diagnose a problem often remains beyond our reach.

Here is a joke about software engineering that you can find in dozens of Internet humor archives. (The Alpine setting might even be an homage to the NATO gathering at Garmisch.)

> A Software Engineer, a Hardware Engineer, and a Departmental Manager were on their way to a meeting in Switzerland. They were driving down a steep mountain

road when suddenly the brakes on their car failed. The car careened almost out of control down the road, bouncing off the crash barriers, until it miraculously ground to a halt scraping along the mountainside. The car's occupants, shaken but unhurt, now had a problem: They were stuck halfway down a mountain in a car with no brakes. What were they to do?

"I know," said the Departmental Manager. "Let's have a meeting, propose a Vision, formulate a Mission Statement, define some Goals, and by a process of Continuous Improvement find a solution to the Critical Problems, and we can be on our way."

"No, no," said the Hardware Engineer. "That will take far too long, and, besides, that method has never worked before. I've got my Swiss Army knife with me, and in no time at all I can strip down the car's braking system, isolate the fault, fix it, and we can be on our way."

"Well," said the Software Engineer, "before we do anything, I think we should push the car back up the road and see if it happens again."

If you report a bug to a programmer, the first thing she will do is ask, "Have you duplicated the problem?"—meaning, can you reliably make it happen again? If the answer is yes, that's more than half the battle. If it is no, most of the time the programmer will simply shrug her shoulders and write it off to faulty hardware or cosmic rays.

"'Software engineering' is something of an oxymoron," L. Peter Deutsch, a software veteran who worked at the fabled Xerox Palo Alto Research Center in the seventies and eighties, has said. "It's very difficult to have real engineering before you have physics, and there isn't anything even close to a physics for software."

Students of other kinds of science are sometimes said to have "physics envy," since, as computing pioneer Alan Kay has put it, physicists "deal with the absolute foundations of the universe, and they do it with serious math." The search for a "physics for software" has become a decades-long quest for software researchers. If we could only discover dependable principles by which software operates, we could transcend the overpowering complexity of today's Rube Goldberg–style programs and engineer our way out of the mire of software time. One day we could find ourselves in a world where software does not need to be programmed at all.

For many on this quest, the common grail has been the idea of "automatic software"—software that nonprogrammers can create, commands in simple English that the computer can be made to understand. This dream was born at the dawn of the computer era when Rear Admiral Grace Hopper and her colleagues invented the compiler. "Hopper believed that programming did not have to be a difficult task," her official Navy biographical sketch reports. "She believed that programs could be written in English and then translated into binary code." Flow-Matic, the language Hopper designed, became one of the roots of Cobol, the business-oriented programming language that was still in wide use at the turn of the millennium.

Hopper's compiler was a vast advance over programming in machine language's binary code of zeros and ones, and it provided a foundation for every software advance to come. But it is no slur on her work to point out that writing in Cobol, or any other programming language since invented, is still a far cry from writing in English. Similarly, the inventors of Fortran, who hoped that their breakthrough language might "virtually eliminate coding," lived to see their handiwork cursed for its crudity and clumsiness.

Yet the dream of virtually eliminating coding has survived in several forms, breeding new acronyms like an alphabetical algae bloom. There is the Unified Modeling Language, or UML, which provides an abstract vocabulary—boxes, blobs, and arrows—for diagramming and designing software independent of specific programming languages. There is Model Driven Development (MDD) and Model Driven Architecture (MDA), terms for approaches to development built around the UML. There is "generative programming," in which the final software product is not written by hand but produced by a generator (which is actually another program). And there is a new buzzphrase, "software factories," which mixes and matches some of these concepts under the metaphoric banner of automating the software building process Henry Ford–style.

The highest-profile effort today to realize a version of the dream of eliminating coding is led by Charles Simonyi, the inventor of modern word processing software at Xerox and Microsoft, whom we have already met as the father of Hungarian notation. For years Simonyi led research at Microsoft into a field he called "intentional programming." In 2002, he left the company that had made him a billionaire and funded a new venture, Intentional Software, with the goal of transforming that research into a real-world product.

Simonyi—a dapper, engaging man whose voice still carries a trace of Hungary, which he fled in his teens—is thinking big. He is as aware of the long, sorry history of software's difficulties as anyone on the planet. "Software as we know it is the bottleneck on the digital horn of plenty," he says. Moore's Law drives computer chips and hardware up an exponential curve of increasing speed and efficiency, but software plods along, unable to keep up.

In Simonyi's view, one source of the trouble is that we expect programmers to do too much. We ask them to be experts at extracting knowledge from nonprogrammers in order to figure out what to build; then we demand that they be experts at communicating those wishes to the computer. "There are two meanings to software design," he says. "One is designing the artifact we're trying to implement. The other is the sheer software engineering to make that artifact come into being. I believe these are two separate roles—the subject matter expert and the software engineer."

Intentional Software intends to "empower" the "subject matter experts"—the people who understand the domain in which the software will be used, whether it is air traffic or health care or adventure gaming. Simonyi wants to give these subject matter experts a set of tools they can use to explain their intentions and needs in a structured way that the computer can understand. Intentional Software's system will let the nonprogrammer experts define a set of problems—for a hospital administration program, say, they might catalog all the "actors," their "roles," tasks that need to be performed, and all other details—in a machine-readable format. That set of definitions, that model, is then fed into a generator program that spits out the end-product software. There is still work for programmers in building the tools for the subject matter experts and in writing the generator. But once that work is done, the nonprogrammers can tinker with their model and make changes in the software without ever needing to "make a humble request to the programmer." Look, Ma, no coding!

In an article describing Intentional's work, technology journalist Claire Tristram offers this analogy: "It's something like an architect being able to draw a blueprint that has the magical property of building the structure it depicts—and even, should the blueprint be amended, rebuilding the structure anew."

Simonyi's dream of automating the complexity out of the software development process is a common one among the field's long-range thinkers. He is unusual in having the singlemindedness and the financial resources to push it toward reality. "Somebody once asked me what language should be used to write a million-line program," Simonyi told an interviewer in 2005. "I asked him how come he has to write a million-line program. Does he have a million-line problem? A million lines is the *Encyclopedia Britannica*—twenty volumes of a thousand pages each. It's almost inconceivable that a business practice or administrative problem would have that much detail in it." Really, Simonyi argues, you should be dealing with maybe only ten thousand lines of "problem description" and another ten thousand lines of code for the generator—a much more manageable twenty thousand total lines containing the "equivalent information content" of that million-line monstrosity.

As a young man, Simonyi led the development of Bravo, the first word processing program that functioned in what programmers now call WYSIWYG fashion (for "what you see is what you get," pronounced "wizzy wig"). Today we take for granted that a document we create on the computer screen, with its type fonts and sizes and images, will look the same when we print it out. But in the mid-1970s, this was a revolution.

Simonyi's Intentional Software is, in a way, an attempt to apply the WYSIWYG principle to the act of programming itself. But Simonyi's enthusiastic descriptions of the brave new software world his invention will shape leave a central question unanswered: Will Intentional Software give the subject matter experts a flexible way to express their needs directly to the machine—or will it demand that nonprogrammer experts submit themselves to the yoke of creating an ultra-detailed, machine-readable model? Is it a leap up the ladder of software evolution or just a fancy way to automate the specification of requirements? Can it help computers understand people better, or will it just force people to communicate more like computers? Simonyi is bullish. But as of this writing, Intentional Software has yet to unveil its products, so it's hard to say.

■ ▋ ▌

Software, we've seen, is a thing of layers, with each layer translating information and processes for the layers above and below. At the bottom of this stack of layers sits the machine with its pure binary ones and zeros. At the top are the human beings, building and using these layers. Simonyi's Intentional Software, at heart, simply proposes one more layer between the machine and us.

Software's layers are its essence, and they are what drive progress in the field, but they have a persistent weakness. They leak. For instance, users of many versions of Microsoft Windows are wearily familiar with the phenomenon of the Blue Screen of Death. You are working away inside some software application like a Web browser or Microsoft Word, and suddenly, out of nowhere, your screen turns blue and you see some white text on it that reads something like this:

```
A fatal exception OE has occurred at
0167:BFF9DFFF. The current application will be
terminated.
```

Eyeing the screen's monochrome look and the blockish typeface, veteran users may sense that they have been flung backward in computer time. Some may even understand that the message's alarming reference to a "fatal exception" means that the program has encountered a bug it cannot recover from and has crashed, or that the cryptic hexadecimal (base 16) numbers describe the exact location in the computer's memory where the crash took place. None of the information is of any value to most users. The comfortable, familiar interface of the application they were using has vanished; a deeper layer of abstraction—in this case, the Windows "shell" or lower-level control program—has erupted like a slanted layer of bedrock poking up through more recent geological strata and into the sunlight. (Perplexing as the Blue Screen of Death is, it actually represented a great advance from earlier versions of Windows since it sometimes allows the user to shut down the offending program and continue working. Before the blue screen, the crash of one Windows program almost always took down the entire machine and all its programs.)

In an essay titled "The Law of Leaky Abstractions," Joel Spolsky wrote,

"All non-trivial abstractions, to some degree, are leaky. Abstractions fail. Sometimes a little, sometimes a lot. There's leakage. Things go wrong." For users this means that sometimes your computer behaves in bizarre, perplexing ways, and sometimes you will want to, as Mitch Kapor said in his Software Design Manifesto, throw it out the window. For programmers it means that new tools and ideas that bundle up some bit of low-level computing complexity and package it in a new, easier-to-manipulate abstraction are great, but only until they break. Then all that hidden complexity leaks back into their work. In theory, the handy new top layer allows programmers to forget about the mess below it; in practice, the programmer still needs to understand that mess, because eventually he is going to land in it. Spolsky wrote:

> Abstractions do not really simplify our lives as much as they were meant to. . . . The law of leaky abstractions means that whenever somebody comes up with a wizzy new code-generation tool that is supposed to make us all ever-so-efficient, you hear a lot of people saying, "Learn how to do it manually first, then use the wizzy tool to save time." Code-generation tools that pretend to abstract out something, like all abstractions, leak, and the only way to deal with the leaks competently is to learn about how the abstractions work and what they are abstracting. So the abstractions save us time working, but they don't save us time learning. . . . And all this means that paradoxically, even as we have higher and higher level programming tools with better and better abstractions, becoming a proficient programmer is getting harder and harder.

So even though "the abstractions we've created over the years do allow us to deal with new orders of complexity in software development that we didn't have to deal with ten or fifteen years ago," and even though these tools "let us get a lot of work done incredibly quickly," Spolsky wrote, "suddenly one day we need to figure out a problem where the abstraction leaked, and it takes two weeks."

The Law of Leaky Abstractions explains why so many programmers I've talked to roll their eyes skeptically when they hear descriptions of Intentional Programming or other similar ideas for transcending software's complexity. It's not that they wouldn't welcome taking another step up the abstraction ladder; but they fear that no matter how high they climb on that ladder, they will always have to run up and down it more than they'd like—and the taller it becomes, the longer the trip.

■ ■ ■

If you talk to programmers long enough about layers of abstraction, you're almost certain to hear the phrase "turtles all the way down." It is a reference to a popular—and apparently apocryphal or at least uncertainly sourced—anecdote that Stephen Hawking used to open his popular book *A Brief History of Time:*

> A well-known scientist (some say it was Bertrand Russell) once gave a public lecture on astronomy. He described how the earth orbits around the sun and how the sun, in turn, orbits around the center of a vast collection of stars called our galaxy. At the end of the lecture, a little old lady at the back of the room got up and said: "What you have told us is rubbish. The world is really a flat plate supported on the back of a giant tortoise." The scientist gave

> a superior smile before replying, "What is the tortoise
> standing on?" "You're very clever, young man, very clever,"
> said the old lady. "But it's turtles all the way down!"

For cosmologists the tale is an amusing way to talk about the problem of ultimate origins. For programmers it has a different spin: It provides a way of talking about systems that are internally coherent. Software products that are, say, "objects all the way down" or "XML all the way down" have a unified nature that doesn't require complex and troublesome internal translation.

But for anyone weaned, as I was, on the collected works of Dr. Seuss, "turtles all the way down" has yet another association: It recalls the classic tale of "Yertle the Turtle," in which a turtle king climbs atop a turtle tower so he can survey the vast extent of his kingdom. Finally, the lowliest turtle in the stack burps, and down comes the king, along with the rest of the turtles in the pile. The moral of *Yertle* is—well, there are plenty of morals. For a programmer the lesson might be that stacks of turtles, or layers of abstractions, don't respond well to the failure of even one small part. They are, to use a word that is very popular among the software world's malcontents, *brittle*. When stressed, they don't bend, they break.

"If builders built houses the way programmers built programs, the first woodpecker to come along would destroy civilization," Gerald Weinberg, the pioneer of computer programmer psychology, once wrote. Leave out a hyphen, the builders of Mariner 1 learned, and your rocket goes haywire.

For Charles Simonyi, the proponents of the UML, and many other software optimists, it is enough to dream of adding another new layer to today's stack of turtles. That's how the field has always advanced in the past, they point out, so why not keep going? But there are other critics of today's software universe whose diagnosis is more sweeping. In their view, the problem is the stack itself. The brittleness comes from our dependence on a pile of unreliable concepts. The entire field of programming took one or more wrong turns in the past. Now, they say, it's time to start over.

■ ■ ■

Alan Kay sits in front of a laptop before a packed crowd. He is showing a slide. The slide presents two photos connected by an ascending curve representing the passage of time. At the curve's end, the top right photo, which shows a collapsed bridge, has a caption that reads, "You finally learn *here* . . ."; at the start of the curve, the bottom left photo of the same bridge being built bears the caption, ". . . what you needed to know *here*."

Kay, one of the central figures in the development of the personal computer, isn't offering a lesson in the power of hindsight. He has named his slide "What is late binding?" and answered the question as follows: "Late binding is being able to make the changes earlier."

Late binding is a term in computer science that refers to programming languages' capacity to provide programmers with more flexibility. Late-bound programs can be changed on the fly, while they're running; you can even build them so that they change *themselves* while they're running. While some of the popular programming languages of today—Java to some degree, Python (Chandler's language) to a greater degree—are considered to have a late-binding nature, the core languages underlying much of the edifice of modern software, C and C++, do not. And none of them, according to Kay, fulfills the promise of true late-binding systems—systems that can be so easily manipulated, even by nonprogrammers, that their flaws are simple to repair and their destiny is to evolve. To find nimbler languages that embody the essence of late-binding dynamic power, Kay says, you have to go back decades—to Lisp, the "list processor" language invented by artificial intelligence researcher John McCarthy in the 1950s, and to Smalltalk, the language Kay himself conceived at Xerox's Palo Alto Research Center in the seventies.

Instead, today's programmers remain in the thrall of the "compile cycle." When they make a change, they have to wait for the system to digest it. First, they run the new code through the compiler, and then they restart the program to see whether the change achieved its goal. Today's faster computers have shortened the compile cycle, but you can still find programmers cursing it everywhere. The Chandler team was no exception. Even though Python offered some of the advantages of the late-binding approach, the tools they were using to build their graphic user interface depended on C++; and even when Chandler itself didn't need to be recompiled, the programmers still faced a cycle of fix-and-restart. You had to wait thirty or even sixty seconds for the program to launch before you could see the results of a change in the code. Those seconds added up to a lot of waiting time.

Kay's modest stature seems inversely related to the broad scope of his ideas, and in recent years he has roamed the terrain of the computing field like a diminutive, mustachioed Jeremiah, denouncing the state of the art, bemoaning the wrong turns the field has taken in the last quarter century, and calling on listeners to forsake their fallen ways and start afresh. "We just don't know how to do software yet," he says. "Computer people are way more self-satisfied with what they have accomplished than they should be. . . . People don't understand that there are alternatives. We need software that's factors of hundreds better than it is."

At Xerox PARC in the seventies, Kay coined the term "object-oriented programming," invented the concept of the overlapping windows interface, and tried to realize his 1968 vision of the Dynabook—an ur-laptop that he dreamed could serve as the ultimate playground for childhood imaginations. In the decades since, he watched the hardware industry gradually deliver computing tools that resembled his ideas—while the software field, in his view, stagnated.

Kay loves to use historical analogies when he talks about software. "If you look at software today," he told an interviewer in 2004, "it's certainly engineering of a sort—but it's the kind of engineering that people without the concept of the arch did. Most software today is very much like an

Egyptian pyramid with millions of bricks piled on top of each other, with no structural integrity, but just done by brute force and thousands of slaves."

You can build big things this way, Kay says, but it "doesn't scale." Ultimately, quantity demands a change in quality; the design that works at one scale fails at another. In a 1997 talk on the twenty-fifth anniversary of Smalltalk, Kay walked through the logic of this argument. "You can make a doghouse out of anything," he declared, but when you enlarge it by a factor of one hundred, you're in trouble: Mass increases exponentially faster than strength.

> When you blow something up by a factor of one hundred, it gets weaker by a factor of one hundred. If you try to build a cathedral that way, it just collapses into a pile of rubble. The most popular response is to say, well that was what we were trying to do all along: put more garbage on it, plaster it over with limestone, and say, yes, we were really trying to do pyramids, not Gothic cathedrals. That I think accounts for much of the structure of modern operating systems today. . . . Or, you can come up with a new concept, which the people who started getting interested in complex structures many years ago did. They called it architecture—literally, the designing and building of successful arches. Non-obvious, non-linear interaction between simple materials that gives you non-obvious synergies and a vast multiplication of materials.

People are amazed, Kay said, when he tells them that the towering spires of Chartres Cathedral required less physical material than the Parthenon. "It's all air and glass, cunningly organized in a beautiful structure to make the whole have much more integrity than any of its parts."

Kay maintains that the software discipline today is actually somewhere in its Middle Ages—"We don't have to build pyramids, we can build Gothic cathedrals, bigger structures with less material"—but that the commercial software world remains stuck in pyramid mode.

He is not suggesting, however, that anyone stop at cathedrals. There can be a real science of computing, he believes; we just haven't discovered it yet. "Someday," he says, "we're going to invent software engineering, and we'll all be better off." And those discoveries and inventions will take their cues, he thinks, from cellular biology. Ultimately, he says, we need to stop writing software and learn how to grow it instead. We think our computing systems are unmanageably complex, but a biologist—who regularly deals with systems that have many orders of magnitude more moving parts— would see them differently, Kay maintains. To such an observer, "Something like a computer could not possibly be regarded as being particularly complex or large or fast. Slow, small, stupid—that's what computers are."

Kay's original vision for object-oriented programming—or OOP, which he pronounces like "oops" without the "s"—was grander than just the idea of organizing code into reusable routines, which the software industry ultimately embraced. Kay-style OOP aimed for a total rethinking of one foundation of our software universe: today's near-universal separation of program from data, procedural code from stored information. Instead, imagine a system in which the information and the code needed to interpret or manipulate it travel together in one bundle, like a cell traveling with its packet of DNA.

Kay first observed such an approach to computing while he was working in the air force as a young programmer in the early 1960s. In a system for transporting data from one Burroughs mainframe computer to another, he noticed that the punch cards interleaved records of stored information with the procedures required to read the information. In the 1970s, at Xerox PARC, he designed Smalltalk around this principle. The name was deliberately modest, "a reaction against the 'Indo-European god theory' where systems were named Zeus, Odin, and Thor, and hardly did anything. I figured that Smalltalk was so innocuous a label that if it ever did anything nice, people would be pleasantly surprised."

In a paper on the history of Smalltalk, Kay wrote,

> In computer terms, Smalltalk is a recursion on the notion of [the] computer itself. Instead of dividing "computer stuff" into things each less strong than the whole—like data structures, procedures, and functions which are the usual paraphernalia of programming languages—each Smalltalk object is a recursion on the entire possibilities of the computer. Thus its semantics are a bit like having thousands and thousands of computers all hooked together by a very fast network.

The result is what Kay calls "a more biological scheme of protected universal cells interacting only through messages that could mimic any desired behavior."

If you are having trouble fathoming this concept, don't feel too bad. You're in the company of several generations of programmers. To Kay's dismay, Smalltalk's most radical innovations never made great headway in the software marketplace, and the kind of object-oriented programming that took over the computing mainstream in the nineties is, in his view, a bastardization. ("I made up the term object-oriented," he says, "and I can tell you, I did not have C++ in mind.") Bundling procedures and data in cell-like portable objects isn't on most programmers' agendas. But for the everyday user there is something seductive about a world in which you would never again have to worry about losing your data when switching from one program or system to another, or fear that a particular document won't make the trip from one program or computer to another. In Kay's world, an object would "travel with all the things it needs."

In the long decades since his breakthrough innovations at PARC, Kay seems to have mostly given up proselytizing among habit-dulled adult programmers and instead focused his energy on demonstrating the power of his ideas to youngsters. Among other things, for the last decade he has labored on Squeak, a latter-day, open source incarnation of Smalltalk for children. Since we are still trying to discover the basics of software engineering, Kay's logic goes, let's give the next generation a taste of alternatives.

In computing, change is inevitable; therefore, according to Kay, the systems we design should let us learn from those changes and leverage them. "At PARC, our idea was, since you never step in the same river twice, the number one thing you want to make the user interface be is a learning environment—something that's explorable in various ways, something that is going to change over the lifetime of the user using this environment." In other words, as users learn new things from the computer, they can change it themselves to take advantage of what they've learned. Like Douglas Engelbart, Kay hoped that computers and their users would together form a positive feedback loop, a bootstrapping of the human brain.

■ ■ ■

Kay is the kind of maverick who has been honored by many in his field but only selectively followed. Yet other central figures in the development of modern software share his complaint that the software profession has taken a fundamental wrong turn. As early as 1978, John Backus, the father of Fortran, was expressing parallel views.

Programming, Backus argued, had grown out of the ideas of John von Neumann, the mathematician who, at the dawn of computing in the 1940s, devised the basic structure of the "stored program" of sequentially executed instructions. But those ideas had become a straitjacket. "Von Neumann languages constantly keep our noses pressed in the dirt of address computation and the separate computation of single words," he wrote. "While it was perhaps natural and inevitable that languages like Fortran and its successors should have developed out of the concept of the von Neumann computer as they did, the fact that such languages have dominated our thinking for twenty years is unfortunate. It is unfortunate because their long-standing familiarity will make it hard for us to understand and adopt new programming styles which one day will offer far greater intellectual and computational power."

Is the entire edifice of software engineering as we know it today a Potemkin village facade? Do we need to start over from the ground up?

Today, one vocal advocate of this view is Jaron Lanier, the computer scientist whose dreadlocked portrait was briefly imprinted on the popular imagination as the guru of virtual reality during that technology's brief craze in the early 1990s. Lanier says that we have fallen into the trap of thinking of arbitrary inventions in computing as "acts of God."

"When you learn about computer science," Lanier said in a 2003 interview, "you learn about the file as if it were an element of nature, like a photon. That's a dangerous mentality. Even if you really can't do anything about it, and you really can't practically write software without files right now, it's still important not to let your brain be bamboozled. You have to remember what's a human invention and what isn't."

The software field feels so much like the movie *Groundhog Day*, Lanier says today—"It's always the same ideas, over and over again"—because we believe the existing framework of computing is the only one possible. The "great shame of computer science" is that, even as hardware speeds up, software fails to improve. Yet programmers have grown complacent, accepting the unsatisfactory present as immutable.

Lanier laid out this critique of the present state of programming in a

2003 essay on the problem of "Gordian software"—software that is like the impossible-to-untie Gordian knot of classical Greece:

> If you look at trends in software, you see a macabre parody of Moore's Law. The expense of giant software projects, the rate at which they fall behind schedule as they expand, the rate at which large projects fail and must be abandoned, and the monetary losses due to unpredicted software problems are all increasing precipitously. Of all the things you can spend a lot of money on, the only things you expect to fail frequently are software and medicine. That's not a coincidence, since they are the two most complex technologies we try to make as a society. Still, the case of software seems somehow less forgivable, because intuitively it seems that as complicated as it's gotten lately, it still exists at a much lower order of tangledness than biology. Since we make it ourselves, we ought to be able to know how to engineer it so it doesn't get quite so confusing.

"Some things in the foundations of computer science are fundamentally askew," Lanier concluded, and proceeded to trace those problems back to "the metaphor of the electrical communications devices that were in use" at the dawn of computing. Those devices all "centered on the sending of signals down wires" or, later, through the ether: telegraph, telephone, radio, and TV. All software systems since, from the first modest machine-language routines to the teeming vastness of today's Internet, have been "simulations of vast tangles of telegraph wires." Signals travel down these wires according to protocols that sender and receiver have agreed upon in advance.

Legend has it that Alexander the Great finally solved the problem of the Gordian knot by whipping out his sword and slicing it in two. Lanier proposed a similarly bold alternative to tangled-wire thinking and spaghetti code. Like Alan Kay, he drew inspiration from the natural world. "If you make a small change to a program, it can result in an enormous change in what the program does. If nature worked that way, the universe would crash all the time." Biological systems are, instead, "error-tolerant" and homeostatic; when you disturb them, they tend to revert to their original state.

Instead of rigid protocols inherited from the telegraph era, Lanier proposed trying to create programs that relate to other programs, and to us, the way our bodies connect with the world. "The world as our nervous systems know it is not based on single point measurements but on surfaces. Put another way, our environment has not necessarily agreed with our bodies in advance on temporal syntax. Our body is a surface that contacts the world on a surface. For instance, our retina sees multiple points of light at once." Why not build software around the same principle of pattern recognition that human beings use to interface with reality? Base it on probability rather than certainty? Have it "try to be an ever better guesser rather than a perfect decoder"?

These ideas have helped the field of robotics make progress in recent times after long years of frustrating failure with the more traditional approach of trying to download perfect models of the world, bit by painful bit, into our machines. "When you de-emphasize protocols and pay attention to patterns on surfaces, you enter into a world of approximation rather than perfection," Lanier wrote. "With protocols you tend to be drawn into all-or-nothing high-wire acts of perfect adherence in at least some aspects of your design. Pattern recognition, in contrast, assumes the constant minor presence of errors and doesn't mind them."

Lanier calls this idea "phenotropic software" (defining it as "the interaction of surfaces"). He readily grants that his vision of programs that essentially "look at each other" is "very different and radical and strange and high-risk." In phenotropic software, the human interface, the way a program communicates with us, would be the same as the interface the program uses to communicate with other programs—"machine and person access components on the same terms"—and that, he admits, looks inefficient at first. But he maintains that it's a better way to "spend the bounty of Moore's Law," to use the extra speed we get each year from the chips that power computers, than the way we spend it now, on bloated, failure-prone programs.

The big problem with software, according to Lanier, is that programmers start by making small programs and learn principles and practices at that level. Then they discover that as those programs scale up to the gargantuan size of today's biggest projects, everything they have learned stops working. "The moment programs grow beyond smallness, their brittleness becomes the most prominent feature, and software engineering becomes Sisyphean."

In Lanier's view, the programming profession is afflicted by a sort of psychological trauma, a collective fall from innocence and grace that each software developer recapitulates as he or she learns the ropes. In the early days of computing, lone innovators could invent whole genres of software with heroic lightning bolts. For his 1963 Ph.D. thesis at MIT, for example, Ivan Sutherland wrote a small (by today's standards) program called Sketchpad—and single-handedly invented the entire field of computer graphics. "Little programs are so easy to write and so joyous," Lanier says. "Wouldn't it be nice to still be in an era when you could write revolutionary little programs? It's painful to give that up. That little drama is repeated again and again for each of us. We relive the whole history of computer science in our own lives and educations. We start off writing little programs—'Hello World,' or whatever it might be. Everyone has a wonderful experience. They start to love computers. They've done some little thing that's just marvelous. How do you get over that first love? It sets the course for the rest of your career. And yet, as you scale up, everything just degenerates."

Lanier's "Gordian Software" article provoked a firestorm of criticism on the Edge.org Web site, John Brockman's salon for future-minded scientists. Daniel Dennett accused Lanier of "getting starry-eyed about a toy model that might—might—scale up and might not." But Lanier is unfazed. He says his critique of software is part of a lifelong research project aimed at harnessing computers to enable people to shape visions for one another, a kind of exchange he calls "post-symbolic communication" and likens to dreaming—a "conscious, waking-state, intentional, shared dream."

His motivation is at once theoretical and personal. In the abstract, he argues, solving software's "scaling problem" could help us solve many other problems. "The fundamental challenge for humanity is understanding complexity. This is the challenge of biology and medicine. It's the challenge in society, economics. It's all the arts, all of our sciences. Whatever intellectual path you go down, you come again and again into a complexity barrier of one sort or another. So this process of understanding software—what's often called the gnarliness of software—is in a sense the principal underlying theoretical approach to this other challenge, which is the most important universal one. So I see this as being very central, cutting very deep—if we can make any progress."

More personally, Lanier's quest is driven by his "disgust" with what he sees as the stagnation of the entire computing field. At a speech at an ACM conference in 2004, he pointed to his laptop as if it had insulted him and—using exactly the same image that Mitch Kapor had two decades before in the "Software Design Manifesto"—exclaimed: "I'm just sick of the stupidity of these things! I want to throw this thing out the window every day! It sucks! I'm driven by annoyance. It's thirty years now. This is ridiculous."

As with most radical new ideas, there is no way to know in advance whether Lanier's phenotropic software will prove a breakthrough or a footnote. Today, outside of a handful of fascinating prototypes, it remains dauntingly abstract. But even as a theoretical outline it helps remind us that the current state of the software art is not etched in stone—that its frustrations are a product of human choices, not laws of nature. And Lanier's diagnosis frames a basic division in thinking about the permanent state of software crisis. Should we "get real" and assume that the only advances in the field will be incremental—modest improvements refining existing models? Or should we be hunting for revolutionary new paradigms, bold strokes that cut the Gordian knot? Should we be tinkering with our software in search of small improvements or eyeing it like Martians, who, as Lanier wrote, might not "be able to distinguish a Macintosh from a space heater"?

These questions are not purely academic in an era when research money is pouring into experimental fields like quantum computing (which substitutes quantum mechanics for today's binary electrical bits, using atoms and photons to store quantum bits, or qubits) and genetic programming (which seeks to "grow" rather than write programs by applying the principles of natural selection to pools of randomly generated solutions to computing problems). But there is no consensus as to whether any such radical innovation will filter down to everyday users and make the miasma of software time a thing of the past in our lifetimes.

Intelligent voices can be found on both sides of this fence. Robert Britcher, in his *The Limits of Software*, takes the pessimistic view to its tragic limit. The past half century's software legacy has become a crushing burden we cannot ever expect to escape, he believes. "We are stuck with the evolutionary pattern created by a hundred, then ten thousand, then millions of computing theoreticians and practitioners and users," he wrote. "There will be no starting over."

Frederick Brooks tells us to give up hunting for a silver bullet: Software's complexity is not a removable quality but an "essential property."

Still, he leaves room for incremental progress made "stepwise, at great effort." Like the medical profession, which sacrificed the prospect of magical tonics when it embraced the germ theory, programmers must abandon hope for cure-alls and accept that "a persistent, unremitting care [has] to be paid to a discipline of cleanliness."

Others hold on to the dream of fundamental reform or revolution. MIT computer scientist Gerald Jay Sussman wrote: "Computer science is in deep trouble. . . . This problem is structural. This is not a complexity problem. It will not be solved by some form of modularity. We need new ideas. We need a new set of engineering principles that can be applied to effectively build flexible, robust, evolvable, and efficient systems."

■ ■ ■

One of the great documents of software history is a brief set of essays by David Lorge Parnas published in 1985. The drab title, "Software Aspects of Strategic Defense Systems," offers no hint of the controversy that birthed it. Parnas, a computer science pioneer and longtime expert on defense software engineering, served on a panel of computer scientists convened to provide advice on the Strategic Defense Initiative (SDI), President Ronald Reagan's proposal for a missile defense program, colloquially known as Star Wars. Parnas had done plenty of work for the military in the past and was no kneejerk peacenik. But in June 1985, he resigned from the committee, declaring that the software required by SDI could never be built.

Star Wars, Parnas argued, was doomed, because neither the software engineering techniques of the time nor those imaginable in the future could ever meet the program's requirements. With devastating precision, Parnas explained why we could not expect to produce a working SDI system whether we relied on conventional software development, with its inefficiency and bugs, or new programming languages and techniques, or any of the silver bullets promised by the software engineering arsenal—automatic programming, artificial intelligence, or mathematical "program verification."

By its very nature, a missile defense program could never be properly tested under actual working conditions; on its very first "field run," it would have to work reliably and effectively to protect millions of lives. No one who understands the nature of programming would ever trust such a system, Parnas wrote. "The easiest way to describe the programming method used in most projects today was given to me by a teacher who was explaining how he teaches programming. 'Think like a computer,' he said. He instructed his students to begin by thinking about what the computer had to do first and to write that down. They would then think about what the computer had to do next and continue in that way until they had described the last thing the computer would do. This, in fact, is the way I was taught to program." The method, Parnas said, works well on small, simple programs. "We think that it works because it worked for the first program that we wrote." Then we introduce complex loops and branches and variables. More ambitious programs involve multiple concurrent processes. "The amount we have to remember grows and grows. . . . Eventually, we make an error."

If "thinking things out in the order that the computer will execute them" is how programmers work, yet doing so is ultimately beyond their capability, how is it that we end up with any working software at all? "Programming is a trial-and-error craft," Parnas wrote. "People write programs without any expectation that they will be right the first time." They make mistakes, and they correct them. They test and fix. They iterate.

People write programs. That statement is worth pausing over. People *write* programs. Despite the field's infatuation with metaphors like architecture and bridge-building and its dabbling in alternative models from biology or physics, the act of programming today remains an act of writing—of typing character after character, word after word, line after line. Tools that let programmers create software by manipulating icons and graphic shapes on screen have a long and sometimes successful history (for a time in the 1990s, one such tool, Visual Basic, won phenomenal popularity). But these have generally served as layers of shortcuts sitting on top of the same old text-based code, and sooner or later, to fix any really hard problems, the programmer would end up elbow-deep in that code anyway.

People write programs.

Bill Joy is one of the leading programmers of his generation; he wrote much of the free Berkeley version of Unix, devised some of the critical underpinnings of the early Internet, and helped create Java. After leaving Sun Microsystems, which he had cofounded, he attempted to write a book. But in 2003 he told the *New York Times* that he was putting the project aside. It was one thing to write for the compiler to interpret or the computer to execute; writing for other people was simply too hard. "With code," he said, "the computer tells you if it understands what you write. It's much harder to write prose. That is, if you want to be understood."

Is programming a kind of creative writing? The notion seems outlandish at first blush. Any discipline that involves complex mathematics and symbolic logic does not seem to share the same cubbyhole with poetry and self-expression. Yet the programming field could learn much from the writing world, argues Richard Gabriel, a veteran of the Lisp and object-oriented programming worlds who is now a Distinguished Engineer at Sun. "My view is that we should train developers the way we train creative people like poets and artists. People may say, 'Well, that sounds really nuts.' But what do people do when they're being trained, for example, to get a master of fine arts in poetry? They study great works of poetry. Do we do that in our software engineering disciplines? No. You don't look at the source code for great pieces of software. Or look at the architecture of great pieces of software. You don't look at their design. You don't study the lives of great software designers. So you don't study the literature of the thing you're trying to build."

One reason for this lies in the computing profession's lack of interest in its own history. But a bigger reason, Gabriel argues, is that much of the software in use today can't be studied; its code is locked away for commercial reasons. (Unsurprisingly, Gabriel is a believer in the open source movement.)

> It is as if all writers had their own private "companies" and only people in the Melville company could read *Moby-Dick* and those in Hemingway's could read *The Sun Also Rises.* Can you imagine developing a rich literature under these circumstances? There could be neither a curriculum in literature nor a way of teaching writing under such conditions. And we expect people to learn to program in exactly this context?

Gabriel's enthusiasm for the notion of programming as creative writing is not purely abstract; in the 1990s, he took three years off from his career to get an MFA in creative writing and to write a poem a day. He discovered that we ask more work of students who want to become writers and poets than of those who aim to become software developers: They must study with mentors, they must present their work for regular criticism by peers in workshops, and they're expected to labor over multiple revisions of the same work. "I think we need to be ashamed of this," Gabriel says. "What we put forward as computer education is a farce."

Gabriel is a radical in his field—"Everything we've done in software for the last fifty years should be literally deleted," he says. Most programmers are not likely to flock to MFA programs. Most probably feel, like Bill Joy, more comfortable writing for an audience of a single computer than for a crowd of human beings.

And yet something extraordinary happened to the software profession over the last decade: Programmers started writing personally, intently, voluminously, pouring out their inspirations and frustrations, their insights and tips and fears and dreams, on Web sites and in blogs. It is a process that began in the earliest days of the Internet, on mailing lists and in newsgroup postings. But it has steadily accelerated and achieved critical mass only since the turn of the millennium and the bursting of the dot-com bubble when a lot of programmers found themselves with extra free time. Not all of this writing is consequential, and not all programmers read it. Yet it is changing the field—creating, if not a canon of great works of software, at least an informal literature around the day-to-day practice of programming. The Web itself has become a distributed version of the vending-machine-lined common room that Gerald Weinberg wrote about in *The Psychology of Computer Programming:* an informal yet essential place for coders to share their knowhow and kibitz. It is also an open forum in which they can continue to ponder, debate, and redefine the nature of the work they do.

■ ■ ■

Engineering or literature? Science or art? Trying to resolve programming's dual-identity crisis has remained a decades-long obsession for many in the field, yet we don't appear to be any closer today than we were when NATO first gathered its great minds in Garmisch and asked them to solve the Software Crisis. Maybe the problem is insoluble. Or maybe it isn't a problem at all but, rather, simply a manifestation of the uniqueness of programming as a human activity.

You can find significant evidence supporting such a conclusion in the life and work of Donald Knuth—programmer, teacher, and author of a series of books that are widely viewed as the bibles of his profession. In these books, Knuth explores the mathematical fundamentals of programming: the essence of algorithms, the abstract sequences of instructions that form the basis for the central activities computers perform when we ask them to sort data or search an array or conduct some other repetitious activity. These volumes represent Knuth's life work. Three have been published since the first came out in 1968, two more are in preparation, and another two are planned. They are the polar opposite of the *Learn This Month's Flavor of Programming Language in 17 Seconds* manuals that the technical books industry churns out. They are not for the faint of heart or mind.

And yet for those adepts who can follow them, these explorations of computing's most mathematically precise, conceptually rarefied regions are anything but arid. They are suffused with humanity. They combine passion for detail and attention to expressive shape in a way that can only be called elegant.

Knuth chose to name his books *The Art of Computer Programming*—not *The Science of Computer Programming*. He has spent a lifetime explaining the title to his colleagues, defining what he means when he calls programming an art. In his 1974 Turing Award lecture, "Computer Programming as an Art," he traced usage of the word *art* back to the late Middle Ages and Renaissance when "*art* meant something devised by man's intellect, as opposed to activities derived from nature or instinct." Assuming a divide between the artifacts of humankind and the products of nature, anything having to do with computing falls unquestionably on the side of art. What could be more artificial than these machines of logic?

Today we're accustomed to distinguishing art not from nature but from science. Knuth readily admits that computing falls on both sides of that line, but it is plain where his passion lies. "The chief goal of my work as educator and author is to help people learn how to write *beautiful programs.* . . . My feeling is that when we prepare a program, it can be like composing poetry or music. . . . Some programs are elegant, some are exquisite, some are sparkling. My claim is that it is possible to write *grand* programs, *noble* programs, truly *magnificent* ones!"

Knuth is, like Frederick Brooks, a religious man (he is also a devotee of the pipe organ). He has pursued the writing and publication of his *Art of Computer Programming* series across five decades with the obsessional purity of someone under the spell of a divine vocation. He is continually revising his volumes to keep up with changes in the field, and he famously offers a reward of $2.56 ("one hexadecimal dollar") to anyone who reports an error in them. He has spent his entire career at Stanford, where he is now a professor emeritus and occasionally takes a break from labors on his book to lecture. But his biography includes one great departure from the ivory tower, a trip into the practical world of programming that derailed his work on *The Art of Computer Programming* for nearly a decade.

In the late 1970s, as Knuth tells the story, he began to notice that the new technical books being published—math and science texts, any books that needed to present complex symbolic notation—looked terrible. The publishing industry was going through its great transition from the era of "hot type" to "cold type," from heavy metal slugs to software-powered type-setting. Large sums were being invested in new basic equipment, and the needs of scientists and mathematicians were not a priority. Because he "couldn't stand to write books that weren't going to look good," Knuth set out to write a software tool for writing and publishing beautiful technical books.

He expected it to take a year. Instead, it took almost ten.

Knuth got his own firsthand experience of software time. Ultimately, he also created programs—the typesetting system TeX and a font design program called Metafont—that became the global standard in their field. Free, and steadily improved and debugged over the years by a large popu-lation of devoted and skilled users, they stand as pioneering examples of the open source approach. Most books you read today that present complex scientific or mathematical symbols are likely to have been written and typeset using Knuth's tools.

Knuth's years-long detour from the airy world of theory into the messy realm of practice left him with an enhanced respect for the obstacles his programming colleagues in the trenches faced. He had always held a skeptical view of his fellow mathematicians' enthusiasm for the idea of finding mathematical proofs that demonstrate programs' "correctness." In one quip that has become a legend in the field, he once wrote to a colleague, "Beware of bugs in the above code; I have only proved it correct, not tried it."

His work on TeX and Metafont led Knuth to draw precisely the opposite conclusion from Bill Joy: Writing software is "much more difficult" than writing books, he declared.

> What were the lessons I learned from so many years of intensive work on the practical problem of setting type by computer? One of the most important lessons, perhaps, is the fact that SOFTWARE IS HARD. From now on I shall have significantly greater respect for every successful software tool that I encounter. During the past decade I was surprised to learn that the writing of programs for TeX and Metafont proved to be much more difficult than all the other things I had done (like proving theorems or writing books). The creation of good software demands a significantly higher standard of accuracy than those other things do, and it requires a longer attention span than other intellectual tasks.

The words come from a series of lectures Knuth has given over the years on "Theory and Practice"; the slide that accompanied this statement presented it in its own illustrative fonts created with his own tools—pillowy-curved lowercase letters for "software" and dark, angular capitals for "HARD."

When he spoke on the subject at a 1986 event at the Franklin Institute commemorating the fortieth anniversary of ENIAC, the first fully programmable electronic computer, Knuth expanded on what he meant:

A longer attention span is needed when you're working on a large computer program than when you're doing other intellectual tasks. I read in *Time* magazine that cots were brought in to the old ENIAC lab so that Eckert [J. Presper Eckert, who oversaw the construction of ENIAC] and others could work through the night. Those of you who slept in those cots will know what I mean. . . . A great deal of technical information must be kept in one's head, all at once, in high-speed random-access memory somewhere in the brain. . . . The amount of technical detail in a large system is one thing that makes programming more demanding than book-writing. Another is that programming demands a significantly higher standard of accuracy.

As a landmark author who also devoted a decade to tackling—and *solving*—a fiendish practical problem in software, Knuth is probably better qualified than any other living human being to compare the relative difficulties of writing books and writing code. His conclusions serve as a gentle warning to anyone who dreams that software can be produced faster or more easily. But, like so many of his colleagues, he is an optimist at heart. He viewed the years spent on TeX and Metafont as worthwhile not only for what they produced but for what they taught him. He even argued that he would make up the lost time on his magnum opus because his new tools allowed him to work more efficiently.

The Art of Computer Programming shows programmers how to design and test algorithms for efficiency. Making some code that sorts information a fraction of a second faster may not seem like a big deal, but if the computer must repeat the algorithm a gazillion times in order to perform some task, your tiny savings can make a huge difference. Incremental fine-tuning at this level requires the same faith that incremental improvement at the level of program design or debugging demands; you have to trust that each little change will contribute to some final value.

Knuth's faith is distilled in the haiku-like poem by the Danish poet/scientist/designer Piet Hein that adorns the entryway to his home:

> The road to wisdom?—Well, it's plain
> and simple to express:
> Err
> and err
> and err again
> but less
> and less
> and less.

■ ■ ■

Don Knuth has made one additional contribution to the programming field that is worth noting here. Although he is best known for his relatively abstruse work in algorithms, he is also the advocate of a concept that he calls "literate programming." Literate programming is intended as an antidote to the excruciating fact that, as Joel Spolsky puts it, "it's harder to read code than to write it." Knuth's proposal emphasizes writing code that is comprehensible to human beings, under the thinking that sooner or later programmers other than the author will need to understand it, and that such code will end up being better structured, too.

> Instead of imagining that our main task is to instruct a
> computer what to do, let us concentrate rather on

explaining to human beings what we want a computer to do. The practitioner of literate programming can be regarded as an essayist, whose main concern is with exposition and excellence of style. Such an author, with thesaurus in hand, chooses the names of variables carefully and explains what each variable means. He or she strives for a program that is comprehensible because its concepts have been introduced in an order that is best for human understanding, using a mixture of formal and informal methods that reinforce each other.

In a sense, the literate programming argument is that if you can fully and clearly document your code, writing it becomes almost trivial. It is an alluring idea, but although it has enthusiasts, it never achieved wide popularity. Most programmers are simply in too much of a rush. Documentation is a distraction.

Besides, they have comments.

Comments are plain English statements embedded between lines of code. Every programming language has a symbol or symbols that programmers use to mark comments that are intended to help other programmers understand the code. The symbols tell the computer, "This isn't for you! Pay no attention to the words behind the curtain!"

Well-commented code is one hallmark of good programming practice; it shows that you care about what you're doing, and it is considerate to those who will come after you to fix your bugs. But comments also serve as a kind of back channel for programmer-to-programmer communication and even occasionally as a competitive arena or an outlet for silliness.

A Norwegian programmer named Vidar Holen lovingly maintains a Web page labeled "Linux kernel swear words." On it he charts over time the levels of profanity in comments within the Linux source code. One graph shows that the absolute levels have risen steadily; another demonstrates that the number of curses *per line of source code* has steadily declined. Is Linux getting more or less foul-mouthed? Holen doesn't try to interpret his statistics.

In 2004, as viruses with names like MyDoom, Bagle, and Netsky began to spread across the Internet to infect Windows computers, researchers discovered that the authors of the viruses were crudely slagging each other in their code's comments: "Don't ruine our bussiness, wanna start a war?" "Skynet AntiVirus–Bagle–you are a looser!!!" This is about as far from Don Knuth as you can get; call it *illiterate* programming.

Comments sometimes serve not just as explanatory notes but as emotional release valves. Despite decades of dabbling with notions of automatic programming and software engineering, making software is still painful. Anguished programmers sometimes just need to say "fuck."

When they do so in commercial code, it is normally under an assumption that the word will remain secret, since no one outside the company will ever see the source. But every now and then we get a glimpse inside this world. In 2004, the source code to parts of one version of Windows 2000 leaked onto the Internet. Fascinated programmers pored through the text. Among the surprises they found were many comments in which the Microsoft programmers berated themselves, their tools, their colleagues, and their products.

```
// around the fucking peice of shit compiler we
pass the last param as an void


// we are such morons. Wiz97 underwent a redesign
between IE4 and IE5


//We have to do this only because Exchange is a
moron.
```

```
// We are morons. We changed the IDeskTray
interface between IE4

// should be fixed in the apps themselves.
Morons!
```

Students of the Windows 2000 source found more than four thousand references to "hacks." In one explosion of typographical attention-getting, a programmer inserted this warning into the code:

```
*!!!!!!!!!!!!!!!!!!!!!!!!!!!!!!!!!!!!!!!!!!!!!!!!!!!!
*!!!!!!!!!!!!!!!!!!!!!!!!!!!!!!!!!!!!!!!!!!!!!!!!!!!!
*IF YOU CHANGE TABS TO SPACES, YOU WILL BE KILLED!!
*!!!!DOING SO FUCKS THE BUILD PROCESS!!!!!!!!!!!!!!
*!!!!!!!!!!!!!!!!!!!!!!!!!!!!!!!!!!!!!!!!!!!!!!!!!!!!
*!!!!!!!!!!!!!!!!!!!!!!!!!!!!!!!!!!!!!!!!!!!!!!!!!!!!
```

Whatever else has changed in the nearly forty years since NATO's best and brightest met to fix the Software Crisis, today's programmers still work in a world where changing tabs to spaces leaves them, basically, fucked. (And not just at Microsoft; Chandler's Python is just as sensitive.) Their comment-tantrums may not be fine art. But given the maddening nature of the problems they still contend with every day, who would begrudge them their outbursts?

CHAPTER 11

THE ROAD TO DOGFOOD

[NOVEMBER 2004–NOVEMBER 2005]

To be effective at any large software project, you have
to become so committed to it. You have to incorporate
so much of it into your brain. I used to dream in code at
night when I was in the middle of some big project.

—*Jaron Lanier*

W hen we left the OSAF developers two chapters ago, in the fall of
2004, they had wrapped up Chandler 0.4 and were beginning
work on 0.5. Once more they set themselves a goal for the cycle: Chandler
0.5 would deliver a "dogfoodable calendar," a shareable calendar that they,
at least, could begin to use. But Chandler had always been envisioned as
a full PIM, a program for managing email and tasks as well as a calendar, and
its design was broader than a plain calendar program would need. The job
of figuring out some design for building a solid working calendar while
leaving room for the program's growth fell to Chandler's designer,
Mimi Yin.

Fitting all of Chandler's functions and ideas onto a single screen was no simple matter, Yin had already discovered, and the goal of removing the silos that segregated different kinds of data only made it harder. Eventually, she believed, Chandler users would rely on her proposed dashboard view— a single screen of items flowing from past to present to future—as the nerve center of their daily information management. But the dashboard had been put on hold; the programmers needed to concentrate on getting the calendar to work. So Yin devised an interim plan for Chandler's interface—one that she hoped would make sense in the current calendar-centric program but also serve it well as its email and task-management functions matured.

On first glance, Chandler's basic three-pane structure resembled that of countless other programs, including Microsoft Outlook. The center of the screen was occupied by the summary view, with its lists of items, and the detail view, with its fields showing information about the item selected in the summary view. Over on the left sat the sidebar, where users would organize their collections of items and select which one to view. Chandler would include several different sorts of collections: "out-of-the-box" collections that the program automatically provided for you, like incoming email, a calendar, and a trash bin; user-defined or "ad hoc collections," which were groupings of items that you created and defined yourself; and shared collections—those that other people had shared with you. The sidebar somehow needed to take all these sorts of collections into account while also adapting to Chandler's silo-free structure, which allowed items to live in multiple collections and users to stamp items, transforming them from one kind into another.

Yin's proposed solution added a row of four buttons to the top of the sidebar's list of collections: All, Mail, Tasks, and Events. This simple button bar (initially called the Kind Filter and later renamed the Application Bar) would serve as a sort of lens or filter controlling which sorts of items showed up in the summary view when you selected a particular collection from the sidebar list. The design's effort to be comprehensive made it powerful but confusing. For instance, there was not only an "All" filter in the button bar, which meant that you weren't filtering out any type of item, but an "All" collection in the sidebar list, which included all the items in your repository. Some use cases represented common circumstances. If, for example, you were looking at your "All" collection and you pressed the "Events" button, you were viewing "all events"—really just a big calendar—and the summary view would switch to a calendar format. Other use cases were more unusual, reflecting Chandler's aspiration toward silolessness. If, say, you had created a collection to contain all the items relating to a home remodeling project, the "All" button would show you the entire collection, but "tasks" and "events" would limit the view to those types of items. Since Chandler's stamping feature allowed you to turn a task into an event and vice versa, some items might show up under both tasks and events.

Yin's plan gestated within OSAF's design team, where it was debated and refined. But when the Apps Group developers had the chance to review it, it caused confusion and consternation. Several of them—particularly John Anderson, who was writing the code to make the sidebar work—simply could not warm to Yin's design. First they found it opaque; later, after weeks of meetings slowly brought its details into focus, they felt it was impractical.

Yin started from ideas about how users would want to organize their workflows; the developers began by imagining the "mental model" the user would develop about the program's functions. The designer argued that the different choices on the Application Bar related to different modes of user behavior. The developers argued that what she really wanted to provide was a set of *views* of the same collection, so why didn't they call it that? Yin saw the Application Bar as a last-ditch compromise, the only remaining piece of an original design that had once included many more ambitious components and capabilities; yet the developers found her sketches overambitious, full of shapes and features that they would have to custom code since there was nothing like them ready-made in their wxWidgets toolkit.

The programmers didn't form a united front against the design. Some objected to it philosophically; others thought it was just not fully thought through; still others thought they might as well give it a try. But their discomfort was obvious, and the meetings on the subject began to bog down in circular conversations.

At the end of one long meeting full of detailed wrangling, John Anderson said he was looking for "more concreteness in the design."

Katie Parlante, whose equanimity was not often ruffled, retorted, "When you say you want 'more concreteness' about the design, you need to be more concrete about what needs to be more concrete!"

By early November 2004, the whole discussion had reached an impasse.

One Tuesday—it happened to be the evening of the presidential election—Yin walked over to Donn Denman's desk and asked to talk. They ended up chatting for more than an hour, trying to bridge the gulf between the worlds of designer and programmer. Denman went home and started to watch the returns, then turned off the TV. ("It was painful for me," he told me. "I grew up in Ohio.") He composed an email to the OSAF development list outlining a compromise that the two had reached. It involved thinking of the intersection of the Application Bar buttons and the collection list as a sort of two-dimensional grid; though no grid would be visible to the user, if you thought about the design this way, if you made it your mental model, it became something a developer could more easily embrace.

Denman sent his email at 12:34 A.M. that night. While it hardly put an end to the issue, it cleared a path toward building a working sidebar for Chandler 0.5.

■ ■ ■

"Designers should all take some time and learn programming," John Anderson declared one day over lunch. "Then they won't keep proposing such really difficult things."

"But then they wouldn't come up with great ideas," Lisa Dusseault replied. "They'd block them. Once you know how to program, you think, it's too hard to do that."

"Well," Anderson said, "I'd still have the ideas. But I'd think about them differently!"

■ ■ ■

In October 2004, Ted Leung posted a note to his blog, which was popular among open source programmers, that OSAF was in the market for a "hot shot Python hacker" to help restructure and refine the Chandler "content model," the definitions the program used to access the data it stored. Improvements here, everyone knew, would go a long way toward speeding Chandler up—helping the program run faster and helping the developers write it faster, too. Phillip J. Eby, a Python programmer who maintained his own blog at dirtsimple.org ("making simple things hard, and complex things impossible," its header joked), answered the call in November. After spending some time studying Chandler's code base, Eby posted to his blog a lengthy entry titled "Python Is Not Java."

```
I was recently looking at the source of a
wxPython-based GUI application. . . . The code was
written by Java developers who are relatively new
```

to Python, and it suffers from some performance
issues (like a 30-second startup time). In
examining the code, I found that they had done
lots of things that make sense in Java, but which
suck terribly in Python. Not because "Python is
slower than Java," but because there are easier
ways to accomplish the same goals in Python that
wouldn't even be possible in Java.

So, the sad thing is that these poor folks worked
much, much harder than they needed to, in order
to produce much more code than they needed to
write, that then performs much more slowly than
the equivalent idiomatic Python would.

There followed a long list of technical recommendations for how to use
Python like a true "Pythonista" rather than a newbie. Eby concluded:

This is only the tip of the iceberg for Java->
Python mindset migration. . . . Essentially, if
you've been using Java for a while and are new to
Python, do not trust your instincts. Your instincts
are tuned to Java, not Python. Take a step back,
and above all, stop writing so much code.

To do this, become more demanding of Python.
Pretend that Python is a magic wand that will
miraculously do whatever you want without you
needing to lift a finger. Ask, "how does Python
already solve my problem?" and "What Python
language feature most resembles my problem?" You
will be absolutely astonished at how often it

> happens that [the] thing you need is already
> there in some form.

Although Eby never mentioned Chandler by name, it was pretty clear who he was talking about. His point was undeniable—that key Chandler developers, no matter how much coding they had under their belts, were Python newbies who simply weren't taking full advantage of the language's features and were sometimes tripping over them instead.

Still, there was something, well, *indelicate* about the way PJE (his customary handle) was publicly exposing the OSAF team's goofs—even as he was interviewing to work with them. Other organizations might have circled the wagons against such behavior. At OSAF, though, Kapor had always encouraged openness to criticism: "If we're anything," he would say, "we're an organization that knows how to learn."

And so Eby came on board in January 2005, working remotely from his home in Palm Beach, Florida, and immediately started drafting a proposal for a massive code cleanup. Some of the work was important but essentially janitorial, involving "flattening the tree"—removing multiple layers of file folder directories in the source code to achieve a simpler, easier-to-grasp structure.

More conceptually ambitious was Eby's idea for rethinking the basic structure of Chandler's code: He devised "a speculative feasibility prototype" named Spike with the goal of "minimizing the number of things somebody has to understand in order to do something useful." Spike would reorganize Chandler's underlying data concepts into five carefully defined and tidily separated layers. That offered OSAF's developers a clean new model, but it also involved clearing the decks in many areas of Chandler and reworking them from scratch—not an inviting prospect after so much time and labor. Parlante and Dusseault agreed not to adopt Spike wholesale but to adapt good ideas from it that could be integrated into Chandler without derailing its slow but steady progress.

Once more Eby turned to his blog to report on his work. He described the process he adopted to create Spike: He had borrowed a method from Extreme Programming called "test-driven development," in which programmers write the test that evaluates the success of a program function *before* they write the function itself. Then, at every stage of work, they can be sure that their code still works and does what was originally intended. With Spike, Eby felt, the result was impressive: "It is quite simply a thing of beauty, an iridescent jewel of interlocking features."

Later, in a postscript, Eby noted that the glowing self-review might sound arrogant. He explained that he had meant to lavish praise on the results of the test-driven method, not to pat himself on the back.

■ ▮ ▯

During 0.4, OSAF had adopted a new approach to sharing in Chandler— abandoning the old peer-to-peer vision and embracing the WebDAV approach Lisa Dusseault had advocated. Morgen Sagen had begun writing code to send items out from Chandler to a WebDAV server and bring them back in again. But what exactly was that server going to be? It turned out that there weren't a lot of mature, widely deployed open source WebDAV servers around, and none that implemented the new calendar-oriented CalDAV standard. There were also a few extra tricks the OSAF developers wanted their server to perform as they tried to preserve some of the special "Chandlerosity" of information that Chandler was going to pass to and from the server.

Dusseault steadily lobbied for the establishment of a separate server project under the OSAF umbrella. It would serve Chandler but exist apart from it. In early 2005, the server project—named Cosmo, after Chandler-the-labradoodle's elder sibling—geared up. It was to be a lean effort, mostly undertaken by a single programmer, Brian Moseley. But because it was starting fresh and because all its dealings with Chandler would transpire using the canonical WebDAV standard, it didn't have to follow any of the Chandler project's precedents. Moseley, for instance, decided to use Java instead of Python.

Before long, Cosmo had its own mailing list, and Dusseault began reorganizing the OSAF wiki so it could accommodate two separate projects. Moseley tended to work at home and rarely showed up for OSAF's meetings. The decision to build a server not only made the whole Chandler effort bigger, as Andi Vajda had predicted, but it also subtly began to fork the OSAF organization. That might have raised alarms if Cosmo hit trouble. But the server moved along without noticeable crisis or delay—a welcome contrast to the many roadblocks Chandler had faced.

It helped that Cosmo employed only one developer; that meant it didn't have to pay a lot of coordination costs. But it was also true that writing server software has always had certain advantages compared with writing software for users. A server is a program that deals almost entirely with other programs and machines; it rarely needs to communicate directly with human beings. And when it does, the human being it needs to talk to—when it is being initially configured, for instance, or when it hits a snag—is usually a pro, a systems administrator or programmer who is already fluent in the server's own dialect.

So for Cosmo there were no encounters with the snakes that hissed at the Chandler team as they tried to reconcile human thought processes and work habits with software systems and interfaces. Moseley and Dusseault faced a relatively manageable set of questions—like, how does a new user set up an account? Or, how do you match up times when a user in one time zone shares a calendar with a user in another? After Chandler's amorphous ambitions and telescoping schedule, Cosmo's carefully defined boundaries came as a relief.

With Chandler 0.5, OSAF's leadership had deliberately scaled back their ambitions and aimed low. They would forget for the moment the promise of organizing the entire universe of data and enabling outside developers to extend the program in unexpected directions. The soul of Agenda was a beautiful thing, they told their team; but for now, could we please just build a working calendar?

Yet not far into the 0.5 schedule, which began at the start of November 2004, it became obvious that even that modest goal was beyond reach. There was simply too far to go to get the application's machinery functioning. There had been some progress: John Anderson and Donn Denman had revamped the CPIA framework according to a new design they called "trees of blocks" that would, they hoped, make future changes faster and easier. David Surovell had finally tamed most of the flicker issues that had plagued Chandler ever since Anderson first logged them as a bug. With the help of WebDAV, sharing was real, and under a new, simple approach to "sharing invitations," Chandler users would simply email a "ticket" to other users that would grant them access to a shared calendar. The mini-calendar navigator (the miniature display of the current month, last month, and next month seen in most calendar programs) now worked well and looked decent. Andi Vajda and Ted Leung had improved the speed of the repository through an accumulation of small and medium tweaks.

But such forward steps did not add up to a calendar that Kapor could begin dogfooding. Meanwhile, an influx of newcomers to the project added more drag to the schedule. With Denman preparing for a paternity leave, OSAF brought in Bryan Stearns, an experienced programmer who had worked with Denman two decades before on the original Macintosh team, and Stearns took over the detail view. Alec Flett, a longtime Netscape developer who had taken some time off programming to work as a school-teacher, and Grant Baillie, a veteran of Steve Jobs's Next who most recently had worked at Apple on its email program, joined in January. The new faces represented reinforcements for the long haul, but in the short term they each had to take some time to figure Chandler out. Brooks's Law had not been repealed.

Chandler was now 1.5 million lines of code, most of which had been incorporated from other projects like wxWidgets and Twisted. There were about 130,000 Chandler-specific lines of Python code that OSAF developers had written. Getting started at the project was, as an OSAF summer intern put it, like moving to a new city and trying to find your bearings.

On March 30, 2005, the day after the 0.5 release, I asked Kapor whether he was at all discouraged by how long a road the project had already traveled and how far it still had to go. "I know we are going to come out with something great," he said. "I just don't know how many more turns of the crank it's going to take."

■ ■ ■

Every March since 2003, Python enthusiasts had gathered in Washington for a conference called PyCon—an idea exchange for dedicated open source programmers. Among other geekish pursuits, PyCon holds "sprints"—quick-immersion workshops in which programmers take on narrowly defined projects and see how far they can get in two or three days.

In 2005, Ted Leung traveled to PyCon with several OSAF developers and helped run a Chandler sprint. They and a group of three interested volunteers produced two separate new extensions to Chandler—parcels that added new capabilities to the program. One of them allowed Chandler to connect to Flickr, the increasingly popular Web-based photo-sharing service; the other hooked up Chandler with another service called Del.icio.us that allowed users to post and share lists of Web bookmarks and to give those bookmarks descriptive metadata tags. With these new parcels a user could import and export photos from Flickr and bookmarks from Del.icio.us, and then organize them from within Chandler. In a similar vein, Chao Lam had run another sprint at a technical meeting of the university consortium that had helped fund OSAF; that produced a parcel which imported people's public "wish lists" from Amazon.com into Chandler.

Once you pulled this stuff into Chandler, you could do Chandler-style things to it, organizing and stamping items, turning them from one kind of thing into another. When Alec Flett demoed the new parcels to the OSAF staff after PyCon, he used the Flickr parcel to import a batch of photos from Kapor's account—all snapshots of his dogs. Then he took one labeled "Cosmo at 3"—a canine birthday portrait, complete with crown—and stamped it as an event for that day. Switching to the calendar view, he showed the photo there on the calendar.

Each extension of Chandler was in itself not a big deal, but as Morgen Sagen's photo album program once did, each breathed life into Chandler's original promise. The calendar was what OSAF's developers had to concentrate on, but it was these little exercises in versatility, these experiments in data alchemy that kept their flames burning and preserved the soul of Agenda across so many delays and setbacks. When Kapor began taking the latest builds of Chandler out to demo for friends in the industry—as he finally did, once Chandler 0.5 was finished—it was these extensions, with their almost effortless stitching together of seemingly disparate types of information and services, that wowed.

Something very unusual had happened to the Chandler team over time. Not by design but maybe not entirely coincidentally, it had become an open source project largely managed by women. Kapor was still the "benevolent dictator for life," the title he had half-jokingly accepted on the Chandler mailing list in a tip of the hat to Linus Torvalds's use of the phrase. But with Katie Parlante and Lisa Dusseault running the engineering groups, Sheila Mooney in charge of product management, and Mimi Yin as the lead designer, Chandler had what was, in the world of software development, an impressive depth of female leadership.

Men have always outnumbered women in software, and the ratio is even more lopsided in open source software. The women who do enter the field often make influential and outsized contributions, but their numbers have never been great, and the figures for women entering computer science programs have actually dropped in recent years. Efforts to explain the disparity risk both invoking stereotypes and profaning sacred cows. But the most thoughtful students of the matter point out that social bias against women tends to trump all other factors. In a 1991 paper, Ellen Spertus, who teaches computer science at Mills College and has studied this issue for years, wrote, "While one cannot rule out the possibility of some innate neurological or psychological differences that would make women less (or more) likely to excel in computer science, I found that the cultural biases against women's pursuing such careers are so large that, even if inherent differences exist, they would not explain the entire gap."

Today, to walk into the management meeting of a software project and encounter a group of female faces is still an exotic experience. No one at OSAF whom I asked had ever before worked on a software team with so many women in charge, and nearly everyone felt that this rare situation might have something to do with the overwhelming civility around the office—the relative rarity of nasty turf wars and rude insult and aggressive ego display. There was conflict, yes, but it was carefully muted. Had Kapor set a different tone for the project that removed common barriers to women advancing? Or had the talented women risen to the top and then created a congenial environment?

Such chicken-egg questions are probably unanswerable. At OSAF, at least, women programmers had a rare shot at running the show. As the show kept growing, one woman in particular found herself handling an unmanageable share of the management duties. Parlante was in charge of long-term architectural issues for Chandler, and she was also the day-to-day manager of the Applications Group, which kept adding new developers. Kapor had decided during 0.4 to hire a new Apps Group manager to report to Parlante, but the search hadn't been easy. It wasn't until the end of 0.5, in March 2005, that the position was finally filled by Philippe Bossut, a methodical French engineer with a wry sense of humor and a thick accent. Bossut had most recently worked at Macromedia, and, like Lisa Dusseault, he had also put in years at Microsoft.

The first thing Bossut did on taking over the apps team reins was to make a spreadsheet listing every single task facing every single developer on 0.6 along with a SWAG (Silly, Wild-Assed Guess) estimating the task's length. Then he would add up all the tasks and see how much time 0.6 would really take. This was different in form but similar in essence to what Dusseault had done with stickies and what Michael Toy had tried to do years before when he asked his developers to estimate their task completion times in Bugzilla. But it went into far greater detail; it was, as the developers put it, "more granular."

"You take a big problem, and you cut it into small problems that you can understand and solve individually," Bossut explained to me. "There is no magic to it. But you have to be thorough."

On his first pass in mid-April, Bossut found that each developer faced roughly forty days of work just to close out the list of bugs that had been left over at the end of 0.5. This meant that 0.6, like nearly all the previous releases, was going to take six months or more, not the three to four months Kapor had originally hoped. "I'm tempted to say I don't care if it takes six months so long as we get the 0.6 we have in mind," he told his managers. The important thing for 0.6, Kapor felt, was to at last devote some time to "fit and finish" work, making the user interface spiffy and snappy. That was the last major obstacle to delivering a calendar that was usable—not necessarily by the early adopters that technology marketers always targeted, but at least by "first adopters," brave souls who most likely would be found within OSAF's own ranks.

And so the developers got to work. They took steady bites out of the bug count. They started adding features to Chandler that were mundane but essential, like color for calendar items, so you could tell where the items came from when you overlaid multiple calendars ("home" and "work," say, or "mine" and "my spouse's"). They reworked the column headers in the table view, which had always looked ugly. They added "busy bars" to the mini-calendar, showing which days had scheduled events. They revamped the entire interface to "internationalize" and "localize" Chandler, allowing the program to be customized in a host of different languages and to support different local customs, like which day of the week comes first. This was the kind of problem that open source made easier to solve: IBM offered a free library that supplied all the specifics you needed about languages and localities, and whether Uruguayans prefer to start their week on Sunday or Monday. It wasn't perfect—in fact, a Uruguayan developer provided some feedback on the Chandler mailing list about some inaccuracies—but it saved mountains of labor.

These were manageable tasks. But there were other, trickier issues dating from Chandler's early days that finally needed to be resolved, too. One was what the developers called the "date/time widget"—the program's device for letting you specify when an event begins and ends. On paper, declaring the date and time of an event is a relatively flexible matter: You can say "5/15 at 3 P.M." or "3 o'clock on May 15" or "15 May @ 3" and the message will get through. Back in the 1980s, Kapor's beloved Lotus Agenda had done a remarkable job of parsing notes for dates and times; it could mine meaning even from vague references to "next Monday." No one expected Chandler to do the same—not yet, anyway. But Mimi Yin hoped at least to overcome some of the most annoying traits of computer-based calendars. She argued that Chandler should model time-date entry on how people speak and proposed that the detail view offer a single field in which users could enter text like "Jan. 3–5, 2004 from 4–6 pm."

"Right there, that's a very complicated message," Bryan Stearns, keeper of the detail view, objected.

Building a smart, people-friendly date/time widget looked like another overambitious schedule buster to the developers. After considerable debate and negotiation, Yin's design argument lost out to their pragmatic objections. For 0.6, each Chandler calendar event would have separate date and time fields for its start and end—four fields in all.

■ ■ ■

If Chandler 0.6 was going to provide first adopters with a calendar they could actually use, another basic feature had to be added: recurring events. These were items you wanted to put on your calendar and then have repeat at regular intervals—such as your company's daily department meeting or your child's weekly guitar lesson. Paper calendars had never provided such a service, but it was exactly what people expected software to do effortlessly.

Recurrence, however, presented Chandler's developers with some practical and philosophical problems. If you wanted to set an event to recur for a specific period—say, every week next month—all was well. But what if you didn't want to specify an end date? What if you wanted the event to repeat indefinitely, until such time as you decided to take it off your calendar?

Software does not handle infinity well. No computer has an infinite amount of storage available to it, and when a program gets caught in an infinite loop, a procedure with no exit, it ceases to respond. Most calendar programs found workarounds to the problem of indefinite recurrence: You could simply store a single "master" version of the recurring event and then display it according to whatever rules governed it. Or you could just choose an arbitrary look-ahead distance of X years and generate the recurrences up to that point and no further. In one other system, Lisa Dusseault reported as OSAF researched the problem, an indefinitely recurring event would replicate itself all the way until 2038, then just stop. (In 2038, it turns out, the internal clocks of all Unix-based computer systems will, like odometers reaching their limit, roll over to zero.) "Apparently," Dusseault wrote in an email, "the world ends in 2038. On the bright side, with no birthdays after 2038 we won't be getting any older."

These tried-and-true workarounds always left some loose ends dangling. For example, when you made a change to a specific instance of a recurring event, like adding notes for that day's meeting, would those changes propagate to each future instance of the event, as you would expect if you had the single-master approach in mind? What, then, if you wanted to add notes to a single instance of a meeting event? While many mature calendar programs had found their own ways to settle this conundrum, each new one to come along had to resolve it all over again.

Chandler's unique design added some further problems to the mix. What happened if you stamped a recurring calendar event as a task? Were you stamping just that one instance or all future instances? If you were looking at events in a summary or list view instead of a calendar format, would you face an endless list of individual instances of indefinitely recurring items? There was trouble here for the user interface, and trouble for the repository on the back end. There were ways of postponing the trouble—for instance, generating specific instances of a recurring event only when the user actually chooses to view a particular time period—but they all harbored gotchas. Grant Baillie pointed out one difficult edge case: "What if a user (possibly through a typo or maybe out of idle curiosity as to what day of the week his/her fortieth birthday will be) jumps to the year 2025—are we going to stuff twenty years' worth of recurring events into the repository?"

In spring and summer of 2005, the OSAF developers' Chandler-dev mailing list began to fill with multiple threads of discussion about how to implement indefinitely recurring events and to limit the difficult implications of their unbounded nature. In one message, Alec Flett wrote, "I think this is actually the hardest problem we're facing in Chandler."

■ ■ ■

A clue to why the deceptively simple-looking matter of event recurrence should prove so difficult can be found in the word *recur* itself. It's not just that some problems recur with such eerie regularity in the making of software that the whole undertaking can feel (as both Mitch Kapor and Jaron Lanier had complained at different times) like the time-travel loop in the movie *Groundhog Day*. It's that recurrence is also another form of the word "recursion," and recursion lies at the heart of everything that is powerful and sometimes untamable about software.

Recurring things repeat, but recursive things repeat in a special way: They loop back into themselves. They are tail-eating snakes.

Mathematically speaking, "recursion" describes a function that calls itself; in computer science, recursion refers to defining a function in terms of itself. The logic of recursion allows simple, compact statements to generate complex and elaborate output. It lies at the heart of many of computing's most celebrated and inspiring systems and ideas. Doug Engelbart's notion of bootstrapping was a recursive technique for accelerating human progress. Lisp—the language invented in the late 1950s by John McCarthy that, decades later, still inspired many of OSAF's programmers—is one great recursion machine. Alan Kay likes to point to McCarthy's "half-page of code at the bottom of page 13 of the Lisp 1.5 manual" and praise it as the "Maxwell's equations of computing"—concentrated, elegant statements that distilled the field's fundamental principles just as James Clerk Maxwell's four equations had laid out the essential workings of electricity and magnetism at the dawn of the machine age. On the page that Kay cited, which provides definitions of two functions named "eval" and "apply," McCarthy essentially described Lisp *in itself.* "This," Kay says, "is the whole world of programming in a few lines that I can put my hand over."

You can put your hand over it, but it is not always so easy to get your head around it. Recursion can make the brain ache. Douglas Hofstadter's classic volume *Gödel, Escher, Bach* is probably the most comprehensive and approachable explanation of the concept available to nonmathematicians. Hofstadter connects the mysterious self-referential effects found in certain realms of mathematics with the infinitely ascending staircases of M. C. Escher's art and with J. S. Bach's playful canons and fugues, and gives all these phenomena a memorable label: *strange loops.* Hofstadter's lucid explorations go on for nearly eight hundred pages of variations on the theme; recursively, *Gödel, Escher, Bach* illustrates its own theme—it loops strangely.

A man on an Escher staircase can climb forever without ever getting any higher. Recursion is dangerous in software because there has to be an exit. Otherwise, your strange loop becomes an infinite loop, and your computer is—to use the precise technical terminology—hosed. As one Lisp textbook put it, "Any recursive function will cause a stack overflow if it does not have a proper termination condition. . . . The termination condition is the most important part of a recursive definition."

Important as it is to make sure that every function or program routine has a "termination condition" and won't just cycle endlessly, when we try to do so, computer science confronts us with a disturbing, unyielding truth: We can't. We can try, but we can't ever be absolutely certain we've succeeded.

At the formative moment of the digital age in the 1930s, as he was defining the fundamental abstractions that would underlie modern computing, the British mathematician Alan Turing devised a surprising proof: He demonstrated that there is no all-purpose, covers-all-the-cases method of examining any given program and its initial input and being certain that the program ever "completes" (or reaches a "termination condition"). Another way to put it is that there's no general algorithm which can prove that any given program will complete for all possible inputs. In other words, there's no program you can write that will assure you that another program will always, eventually, come to a stop no matter what data you give it.

As David Harel explains in his book *Computers Ltd.*, "It is tempting to try and solve the problem by a simulation algorithm that simply mimics running the program . . . and waits to see what happens. The point is that if and when execution terminates we can justifiably stop and conclude that the answer is 'Yes' [the program does complete]. . . . The difficulty is in deciding when to stop waiting and say 'No.' We cannot simply give up after a long wait and conclude that since the simulation has not yet terminated it never will." Answering this question would require infinite time; since neither we human beings nor the fastest computers in the world have that luxury, the question is undecidable.

No one has since been able to disprove Turing's counterintuitive but logically demonstrable theory. It is known as "the halting problem." Sitting at the outer boundaries separating what can and cannot be computed, it represented the first in what has turned out to be a surprisingly large set of abstract problems that computers, for all their mind-boggling speed and power, simply cannot solve.

Conundrums like the halting problem and other similarly undecidable problems that haunt the sleep of computer scientists have never stood in the way of writing everyday software—the programs we use to track profits and losses, keep a calendar, or write books. But they have real, practical consequences for efforts to improve software. It turns out that the dream of an automatic program "verifier"—a program that could examine any other program and determine whether that program provides the correct output for any given input—is doomed to remain a dream. The question is undecidable. Your program might provide accurate verification for many individual cases, but you could never be absolutely certain it was right in any particular case. As Harel puts it, this "dashes our hope for a software system that would make sure that our computers do what we want them to do." Such a system is not simply difficult to imagine, hard to design, and elusive to implement; it is, to the best of our mathematical understanding today, fundamentally impossible.

■ ■ ■

The halting problem and similar undecidables are matters of formal logic and demand definition through symbolic notation. We lose precision by explaining them in what programmers call natural language, as I've struggled to do here. But we gain something, too. They are also useful metaphors.

As OSAF's developers grappled with how to implement recurring events in Chandler, I found myself facing my own author's version of the halting problem. I had been following the labors of Kapor and his programmers for nearly three years. When I first showed up at OSAF in January 2003, they believed that a finished Chandler—or at least a usable product that could be labeled 1.0—was a year away. *Sure*, I thought. I'd been around the block once or twice; I immediately doubled that number as I planned my book's schedule. I would do two years' research, I figured. Then Chandler would be done, or close to it, and I would write.

But my skepticism was inadequate. The Chandler schedule had stretched out like a shaggy-dog story. Deep into year three there was still no "termination condition" in sight. At what point would the story reach a natural conclusion? Worse, could I ever know for certain that it *would* reach a conclusion?

In planning my project I had failed to take into account Hofstadter's Law, the recursive principle to which Douglas Hofstadter attached his name: *It always takes longer than you expect, even when you take into account Hofstadter's Law.*

This strange loop seemed to define the essence of software time. Now I was stuck in it myself.

After three years, Chandler was beginning to become a somewhat usable, though incomplete, calendar program. But I could not say with any confidence how much longer it would take for the project to deliver something like its original promise. OSAF's managers had a plan that involved four more "dot releases" over roughly two more years to arrive at a 1.0 version of the program—but the shape and scope of that program remained open to further revision and course correction.

It seemed that ever since I had started visiting OSAF's programmers, their target held at a constant distance of roughly two years; as real time passed, the goal receded like a fugitive horizon. Their work had helped me find some answers to the questions about software that I started out with. But now my story's threads were beginning to vanish into a software time black hole.

■ ■ ■

Like so many problems in writing software, the question of how to handle recurring events in Chandler proved susceptible to a methodical campaign of divide, conquer, and punt: Divide the problem into manageable parts, conquer enough of them to make things work okay, and postpone tackling the most difficult pieces for the indeterminate future. The developer who ended up tackling most of the recurrence problems, Jeffrey Harris, had started out as a long-distance participant in the project, posting to the mailing lists and contributing code patches. Harris had studied physics at the University of Michigan, then joined an "intentional community" in Missouri called Dancing Rabbit Ecovillage, where he built houses, sang, and ran a software consulting business.

In summer 2005, he moved to San Francisco to join OSAF full-time, becoming one of a growing number of OSAF employees who had first hooked up with the project from a distance or as volunteers. Mike "Codebear" Taylor lived in Philadelphia and had been contributing comments and ideas on Chandler for a couple of years when OSAF asked him to take over the tricky job of managing its builds—tending the source tree, overseeing the mechanics of each milestone release, and making sure that when the Tinderbox charts turned red and the build broke, it got fixed fast.

That such a central operational role could be filled by someone on the other side of the continent from the main office was one sign that OSAF had moved a little further in the direction of behaving like a "real" open source project. In 2005, OSAF's developers also moved more of their technical debates out of the meeting rooms and private email exchanges and onto the project's public mailing lists so that volunteers and outsiders could participate on a more equal footing.

But writing open source code for three years was not the same thing as building an open source community. As Chandler 0.6 neared completion, Ted Leung sent Mitch Kapor a report assessing OSAF's successes and failures in its open source efforts. He found that of approximately 4,400 total bugs logged in Bugzilla to date, around 100 had been filed by people outside OSAF. And there had been only a handful of actual code contributions from outsiders. Kapor and his team had established some of the necessary preconditions for a successful open source project, like publishing their code under an open source license and making their research and deliberations public on the Net. They were nothing if not transparent. But, Leung wrote, "there's a difference between transparency aimed at giving visibility and transparency that is aimed at producing collaboration."

OSAF still had a long way to go to achieve the latter. The Chandler bazaar still lacked tents. And being open source hadn't prevented Chandler from ending up in the same agonizing time warp as countless other ambitious software projects. Maybe Eric Raymond's "The Cathedral and the Bazaar" had been wrong, and Linus's Law ("'Given enough eyeballs, all bugs are shallow") didn't transcend Brooks's Law after all. Or perhaps OSAF, for all its transparency, had so far failed to meet Raymond's requirement for success with a bazaar-style open source project—that it must recognize, embrace, and reward good ideas from outsiders.

Richard Stallman, the godfather of free software, liked to say, "When people ask me when something will be finished, I respond, 'It will be ready sooner if you help.'" OSAF welcomed volunteers and external contributions, but Chandler's grand design ambitions and sluggish pace of delivery had made it hard for outsiders to pitch in.

In "The Cathedral and the Bazaar," Raymond wrote,

> It's fairly clear that one cannot code from the ground up in bazaar style. One can test, debug and improve in bazaar style, but it would be very hard to originate a project in bazaar mode. . . . Your nascent developer community needs to have something runnable and testable to play with. When you start community-building, what you need to be able to present is a plausible promise. Your program doesn't have to work particularly well. It can be crude, buggy, incomplete, and poorly documented. What it must not fail to do is convince potential co-developers that it can be evolved into something really neat in the foreseeable future.

As I talked to people outside OSAF about Chandler, the most common criticism ran something like this: Kapor had missed the boat by not building Chandler inside the browser from the start. That would have saved all the extra work of building cross-platform software for three different operating systems, and also all the labor of building the complex wxWidgets-based graphic interface. The Web browser takes care of all that for you. It already runs on multiple platforms and provides a ready-made user interface. Now, people were saying in 2005, it was obvious that Web-based software was the future. The Gmail-style AJAX tricks were hot. Firefox was hot. The browser was the place to be.

Kapor was stoic about this line of criticism. First of all, he would point out, the world had looked different in 2002 when Chandler's genetic structure was first set. Furthermore, Web-based software did you no good if you weren't able to work online (though in the United States, at least, the wide availability of broadband and wireless connections meant that for many people the only place they were likely to spend extended time offline was the interior of an airplane). Finally, the Web-based AJAX-style approach, exciting as it was, hadn't yet proved itself as a vehicle for the kind of novel, flexible design Chandler aimed for.

Still, Kapor admitted on his blog that he was beginning to see things differently. Even this self-described "old applications guy" was now on the verge of accepting that Web-delivered, browser-based software had begun to overtake the desktop-based programs of personal computing's golden age. From now on, he declared, he would most likely adopt a Web interface for any new project from the start.

In the summer of 2005, as work on Chandler 0.6 and the early versions of the Cosmo server progressed, OSAF began a third small project: a Web-based interface, or client, for users to access any calendars they might have stored on a Cosmo server. The idea was to serve not only Chandler users but users of other calendar programs—including the Macintosh's iCal and any other calendar software that supported the new CalDAV standard.

Kapor often said that one reason OSAF ought to focus on building a good calendar was that people really needed one. Based on the number of small start-up companies and software projects that emerged during 2005 to offer one or another kind of calendar service, he was right. But the more new calendars that emerged, the more of a mess users and programmers would face if these multiple calendars weren't compatible with one another. It was already a "train wreck," wrote *Infoworld* columnist Jon Udell. Software writer Scott Mace started a blog tracking the problem; he titled it Calendar Swamp.

OSAF's developers hoped that their CalDAV standard could tie together the loose calendars of the world and help drain the swamp. Lisa Dusseault and Brian Moseley began hunting for two or three more developers to build the new Web-based client. In keeping with OSAF's dog-based product-naming scheme, they named it Scooby.

■ ■ ■

The original release date for Chandler 0.6 was September 2005. By late summer the mountain of bugs still looked overwhelming, and the managers agreed to let the deadline slip to the end of November. This time they were determined to produce dogfood. They would release a program that could in some provisional way actually be used. And that, they knew, would change the world for them: Instead of guessing what to do next, they would hear comments and complaints from real users. It didn't matter whether those real users were halfway around the world or sitting in the conference room next to them; the important thing was to get some hands-on feedback.

But something unexpected happened as 0.6 moved from milestone to milestone. People at OSAF started dogfooding Chandler early. Something was happening *ahead* of schedule. For once, software time had run faster than real time.

On September 23, 2005, Philippe Bossut began using Chandler as his personal calendar. Almost immediately he started finding new bugs, particularly with recurring events. He blogged about his experience: "I ran through a couple of crashers that I carefully walk around in my daily usage. I'm avoiding anything too adventurous with recurring events (like sharing or moving between collections . . .). But despite the bugs, the great news is: Chandler is rapidly becoming usable; there's nothing I was doing with iCal I found myself not being able to do in Chandler."

Soon after, Sheila Mooney started dogfooding Chandler, too. Her husband had taken a new job and was about to move to Vancouver, where she planned to work for one week each month. The long-distance relationship would give her some good opportunities for testing the calendar's sharing features.

By the end of October, Kapor himself was on the verge of dogfooding— importing his existing iCal events into Chandler and using it as his day-to-day calendar. One thing stood in the way: Chandler's recurring events feature allowed you to repeat events daily, weekly, or monthly. But Kapor had several items on his calendar that needed to recur biweekly.

By this point in the 0.6 development cycle, the release was considered "feature-complete"; you weren't supposed to add new features, just concentrate on fixing the highest-priority bugs. But the developers slipped in one little extra under the wire: Chandler 0.6 would allow you to create biweekly recurring events.

■ ■ ■

The halting problem is not just a theoretical enigma. The participants in every software project run smack into a real-world version of it as they begin to search for an end point to their work. And each team relearns the same painful lesson: There is no foolproof procedure for telling when a project is done and whether it has achieved its goals.

In 1999, Alan Cooper—a software developer who created much of the Visual Basic programming language in the early nineties and is now a prominent software design advocate—published a fierce book titled *The Inmates Are Running the Asylum* that provides a rap sheet of the software industry's sins. In it Cooper wrote, "Software development lacks one key element—an understanding of what it means to be 'Done.'"

This halting problem has practical consequences for software development schedules—consequences that Chandler's developers had struggled with for three years. When software arises in the crucible of a start-up company responsible to investors, dwindling cash accounts and the demands of the cash's suppliers typically enforce a project's "termination condition." Alternately, pure open source projects simply move at a pace governed by the waxing and waning of volunteers' energy. In Linus Torvalds's words, "Because the software is free, there is no pressure to release it before it is really ready just to achieve some sales target. Every version of Linux is declared to be finished only when it is actually finished, which explains why it is so solid." Chandler sat at an unusual midpoint between these approaches, leaving it without clear signposts to mark the moment of "done" or even to indicate the direction in which completion might lie.

But there is another consequence of software development's halting problem, one that is less pragmatic than existential. David Allen, the *Getting Things Done* guru, talked about the "gnawing sense of anxiety" suffered by knowledge workers who face mountains of open-ended tasks. Software developers always have more to do; the definition of "done," even for an interim release or small milestone, is always somewhat arbitrary. In this their work is more like an author's than a builder's. "Done" isn't something that is obvious to an observer. "Done" is something you must decide for yourself.

Chandler, obviously, was not done, not even remotely. Still, with people using it to schedule their days, it was real and alive in a way it had never been before. Its developers had reached a psychological milestone; they could put at least a chunk of that "gnawing sense of anxiety" behind them.

■ ▌ ▌

When software *is*, one way or another, done, it can have astonishing longevity. Though it may take forever to build a good program, a good program will sometimes last almost as long; this is yet another dimension of software time. Programs written in Cobol, the mainstream business programming language dominant in the 1960s, were still in wide use in the late 1990s. To conserve once-scarce space in memory, many of them had allowed only two digits for the year field. As 2000—Y2K—loomed, they had to be fixed or decommissioned.

More recent software history provides a wealth of examples of programs that flourished after hard times or returned to life after consignment to oblivion. On Technology, the company Mitch Kapor founded after Lotus, produced only a fraction of the software that Kapor had hoped it would, but its MeetingMaker group scheduling program has remained in wide use for two decades. Next, the ambitious company that Steve Jobs started after he was ousted from Apple in 1985, failed to change the world or even achieve any significant adoption by businesses or consumers. But the Next system lives on as the foundation for today's Macintosh OSX operating system.

Remember that little content management system my colleagues at Salon built in 2000, the one whose disastrous deployment haunted me and launched me on this book's inquiry? It turned into an open source project named Bricolage and has been used by the World Health Organization, the RAND Corporation, and the Howard Dean presidential campaign.

And all the research for the book you're reading was organized, compiled, outlined, cross-referenced, and endlessly tweaked in an old Windows program called Ecco Pro that was first released in 1993. In 1997, Ecco was "orphaned" by its owners, who felt they couldn't compete with Microsoft Outlook (though it makes Outlook's rigid PIM design look like a—Oh, never mind!), and the Ecco code has not been touched in nearly a decade. But it runs beautifully on my Windows computer, and it has never lost a single piece of data I have entrusted to it in all the years I've used it. It is not cross-platform or open source or Web-based. But it still does certain kinds of outlining and flexible organizing (the kind of thing Chandler promises but hasn't yet delivered) better than any other program I've found. Among the programmers I got to know at OSAF, I was happy to find several fellow Ecco enthusiasts.

Old code rarely offers trendy graphics or flavor-of-the-month features, but it has one considerable advantage: It tends to work. A program that has been well used is like an old garden that has been well tended or a vintage guitar that has been well played: Its rough edges have been filed away, its bugs have been found and fixed, and its performance is a known and valuable quantity.

■ ▌ ▎

As I write in mid-2006, Chandler's future life—millions of users? a footnote?—is impossible to forecast. All we can bet on is that its story is likely to continue to take unpredictable turns. Ending up somewhere entirely different from where you expected to go is the norm in this world. Software projects are prime illustrations of the law of unintended consequences, and their innovations and breakthroughs are more often side effects than planned outcomes.

Ward Cunningham's wiki, for example, came about not because the programmer set out to build Web pages that anyone could add to but because he wanted his collaborators to be able to fix their own mistakes. The young developers who started a small company named Pyra in 1999 aimed to produce a collaborative project management tool; they also built a small program to write a Web journal and communicate with their users. That little side project became Blogger, the largest blogging service in the world, with millions of users (Google acquired it in 2003). Ludicorp was a software company that prototyped a system for a "massive multiplayer online role-playing game"; the project, called the Game Neverending, ended pretty quickly, but its creators revamped some of its parts and built the Flickr photo-sharing service, which became the toast of the Web in 2004 (and was acquired by Yahoo! in 2005).

For that matter, who could see, looking at a little effort by a programmer at a European particle-physics lab to ease the sharing of research on the Internet in 1991, that it would evolve into today's World Wide Web?

Some observers already credit Chandler with one major unintended consequence. According to this view, it makes little difference whether Chandler ever becomes a useful program or achieves any of its original goals; what matters is that in building his open source organization, Kapor got to know Mozilla's Mitchell Baker, and when AOL/Netscape decided to stop supporting the open source browser, he was in the right place at the right time to help rescue it, midwife the birth of the Mozilla Foundation, and set Firefox on its successful trajectory. As I write, Firefox has come from nowhere to win more than 10 percent of the browser market (more than 212 million downloads as this book goes to press). Its competition has forced Microsoft to revisit its long-fallow Internet Explorer product and get back into the business of browser development, with a payoff for Web users everywhere.

When Chandler started out, this was not a chapter Kapor or anyone else could have foreseen. There will certainly be more such chapters.

■ ■ ■

As Chandler's developers prepared to release Chandler 0.6 in November 2005 and to begin its "dogfooding" in earnest, they had already celebrated four births and three marriages. They had welcomed dozens of new colleagues and said good-bye to nearly as many. They had written tens of thousands of lines of code. They had logged thousands of bugs and fixed a fair portion of them. They also faced a software market that had changed radically since they first dreamed up their project.

Microsoft's new version of Windows was beginning to lumber down the pike, plagued by further delays that pushed its planned release date back to early 2007. It had dropped the Longhorn code name and was now to be known as Vista—ironically, the same name Andy Hertzfeld had chosen for Chandler's early prototype. Windows Vista had once promised a Chandler-like revamping of computer users' relationship with their personal information; now, despite the half decade it took Microsoft to create it, it seemed likely to be a much more modest upgrade. But while Microsoft had dallied and Chandler had struggled to find its footing, the Web-based software ecosystem had exploded: In 2005, dozens of small new ventures began competing to offer Internet users new personal information management services—including a half-dozen Web-based calendar services alone.

I talked with Kapor near the end of the 0.6 release cycle and wondered whether he found this landscape disheartening. Could Chandler really set itself apart? Absolutely, he replied: 0.6 already had a handful of potential "killer features." The most important was simply read/write sharing: giving people the ability to publish a calendar so that other people could read *and edit* it. Chandler's approach to data added some intriguing wrinkles to sharing: In the case of, say, two busy spouses sharing each other's calendars, as Kapor was beginning to do with his wife, if you took an item on your calendar and added it to your spouse's, then later changed that item on your calendar, the changes would propagate over and stay in sync on both calendars. "Nobody does that!" Kapor exclaimed.

He was right. This was, as he put it, "actually very cool." But it was still a long way away from the world-changing ambition of Chandler's early days, and Kapor admitted to feeling "humbled."

I asked him what advice he would give himself if he could time-travel back to the start of the project. "It's a twenty-person project," he replied. "It's not a one- or two- or three-person project, and it's not three hundred people. The ambition requires a medium-size team, and when you have a medium-size team, that comes with consequences. There's a lot of coordination. We're still learning, and we're going to continue to learn. How do you execute? How do you set goals? Because I had not done this kind of hands-on twenty-person project with this scope of ambition before, there are so many things that have been surprising and continue to be a surprise.

"We've consistently overinvested in infrastructure and design, the fruits of which won't be realized in the next development cycle or even two—that is, not in the next six or twelve months. You pay a price for that in a loss of agility. The advice I would give is to do even more of what we've been doing in the last couple of years, which is to sequence the innovation, stage things, and be less ambitious. Do not build out infrastructure, like CPIA, except insofar as you need to meet the goals of the next year. I'm more and more feeling like the art here is to do agile development without losing the long-term vision—and, frankly, I didn't even define the problem as that to start with."

With all the dispiriting delays, had he ever considered shutting Chandler down? "No. There were times when I felt horribly depressed.

But I've learned in my life, not just here, that if I have very powerful feelings of hopelessness, I should sit with them. I should refrain from taking action—because those feelings tend to be transient; they tend to be triggered by circumstances. Instead, just personally, take some time out, whether it's an hour or a few hours or a day or two. And that's as long as it's ever been with this process to regain perspective. And every time I've come back, saying I believe we can find a way to accomplish the long-term vision by adapting how we go about doing it."

OSAF, I said, finally seemed to accept that its release cycle was six months long; earlier in the project they had kept aiming for much shorter deadlines.

"There was more wishfulness—a lot of wishfulness," he said wistfully. "I think it just doesn't help to be unrealistic. If you can find an effective way to deal with the real circumstances without becoming demoralized or breaking apart, it's better to go through the pain of incorporating reality than not. So we've just continued doing it, and we are not done with it. Developing software is still hard. It's not all that much easier twenty years later than when I first started. You can do more, but you have fundamentally similar problems of coordination. That's the universe we're living in now, so we have to make appropriate adjustments—cut our food into smaller pieces and chew it in smaller bites. And digest. And move on.

"I do think," he added, "that organizationally and personally we've learned an enormous amount about how to develop software. We've kind of reinvented the wheel. These are things that other people know, so it's taken us a little longer to learn those things. But having learned them, that's a kind of intellectual capital, and I'm absolutely firmly intent on reapplying it and staying the course."

| | | |

Chandler started with a vision—the soul of Agenda reborn in a contemporary program, the leveling of silos, and the empowering of users to slice and dice data any way they choose. That vision had a simple elegant heart. But software development takes simple elegant visions and atomizes them, separating them into millions of implementation details and interface choices and compromises. The art of making software well is, in a sense, the ability to send a vision through that atomizer in such a way that it can eventually be put back together—like the packets of data that separate to cross the Internet and get reassembled into coherent messages for your computer. Or like a Starship Enterprise crew member dissolved into particles in the transporter room and shimmered back into human shape on a planet's surface.

In late 2005, Chandler had passed through that beam and was beginning to glimmer into view. Most of the software enthusiasts and open source devotees who had flocked to the Chandler mailing lists when Kapor first announced OSAF three years before had moved on, and many had written Chandler off as a dead end or a lost cause. Would Chandler fall by the wayside, trampled under by the roaring AJAX crowd? Or would it, like Mozilla, after years in the wilderness, emerge to win the hearts and desktops of a multitude?

I had a selfish interest in Chandler's success. Someday, after all, after some operating system upgrade or other random change in the computing environment, my Ecco program would stop working. I was just starting to play around with using Chandler as my calendar; by the time I was ready to write another book, it would be great to use Chandler to organize my research. In the meantime, though, my Chandler calendar had an important date looming on it.

It was time to bid OSAF and Chandler farewell, and not just because I had a manuscript to turn in. Kapor and his team had learned a great deal about why making software is so hard and how to make things a little easier. So, I hope, have we. But I must leave the Chandler story itself unfinished, open-ended—because, for better or worse, it *is* open-ended.

By the time this book is published, OSAF will have released perhaps two more versions of the program. You can go to the Web site now and, if you're a programmer, download the source. Or maybe you'll just want to run the "end-user installer" and see how much further the project has traveled. Has it gotten trapped in yet another loop of software time? Or has it achieved the "plausible promise" of something great, inspiring a community of developers to kick into gear and make it greater?

Here's one data point: While I was writing this chapter, an icon alerted me to some new incoming email. I clicked over to my email program and saw it was a posting to the Chandler-dev list. Davor Cubranic, a programmer at the University of Victoria in British Columbia who had been hanging out on the Chandler mailing lists, presented a suggestion: "I'm a big fan of Ecco PIM/outliner, and one of the very useful features in its UI is the way it allows selecting time by clicking in an analog clock dropdown [a menu shaped like a clock face]. . . . So having to enter times in Chandler by typing them in a text control felt like a step backward, and I tried to recreate Ecco's clock dropdown in wxPython."

Cubranic's email didn't just propose the idea; it contained about 180 lines of Python code that implemented it. "The code has been tested on Windows and wxPython 2.6.1.0," he wrote.

Ted Leung wrote back, "Thanks for the code! I tried running it on Mac and Linux. It works fine on Linux." But on the Mac it caused an error.

Alec Flett responded enthusiastically, too, but said that it was too late in the 0.6 cycle to add a new feature.

Heikki Toivonen suggested that Cubranic file it as a bug in Bugzilla, which had a special category for things that were really "enhancements" more than bugs.

So Cubranic entered it as bug 4520: "An alternate way to enter time in event detail view."

A LONG BET

Somewhere, someone's fist is pounding the table again. Why can't we build software the way we build bridges?

Well, maybe we already do.

As OSAF's programmers labored to construct their tower of code, piling bug fixes atop Python objects atop the data repository, I watched the work proceed on the new eastern span of the Bay Bridge. The project, replacing half of the 4.5-mile, double-decker bridge that several hundred thousand vehicles cross each day, was born in the 1990s and first budgeted at a little over a billion dollars. The original design called for a low-slung, unadorned causeway, but political rivalries and local pride led to the adoption of a more ambitious and unique design. The new span, a "self-anchored suspension bridge," would hang from a single tower. A web of cables would stretch down from that lone spire, underneath the roadway and back up to the tower top, in a picturesque array of gargantuan loops. It was going to be not only a beautiful bridge to look at but a conceptually daring bridge, a bootstrapped bridge—a self-referential bridge to warm Douglas Hofstadter's heart.

There was only one problem: Nothing like it had ever been built before. And nobody was eager to tackle it. When the State of California put it out to bid, the lone contractor to throw its hat in the ring came in much higher than expected.

In December 2004, California governor Arnold Schwarzenegger stepped in and suspended the project, declaring that the Bay Area region would have to shoulder more of the ballooning cost of the project and calling for a second look at the bridge design. Never mind that work on half of the bridge, the water-hugging viaduct that would carry motorists for more than a mile on a slow climb up to the future main span, was already very far along, and every morning you could see vehicles swarming up a temporary ramp onto the new roadbed. Schwarzenegger wanted the project scaled back to a less novel and cheaper design. The governor, the state legislature, the state's transportation agency, and local governments spent months bickering and horse-trading. The transportation agency claimed that each day of delay was costing the state $400,000. Finally, in July 2005, a new compromise reaffirmed the fancier single-tower design, to be paid for with bridge toll hikes and other measures, and projected a new finish date for the bridge: 2012—almost a quarter century after the Loma Prieta earthquake had shaken a chunk of the old bridge deck loose.

As I read about the controversy, I couldn't help thinking of all the software management manuals that used the rigorous procedures and time-tested standards of civil engineering as a cudgel to whack the fickle dreamers of the programming profession over the head. "Software development needs more discipline," they would say. "Nobody ever·tried to change the design of a bridge after it was already half-built!"

The State of California had done a fine job of undermining *that* argument.

▮ ▮ ▮

Why can't we build software the way we build bridges? Not, of course, the way California is building bridges; people who ask this question are dreaming of a rigorous discipline founded on reliable formulas. But until and unless we can devise a "physics of software" to match the calculations of mass and motion and energy that govern engineering in the physical world, we probably can't do that. Nonetheless, the technical difficulties of software development have grudgingly yielded to incremental improvement. The same thing happened in the world of bridge building. Today, our competence in this realm is something we take for granted; we have forgotten (or repressed) the long history of bridge failures and collapses through the nineteenth century and into the early twentieth, when new technologies and immature practices created a record of disaster and loss that in many ways dwarfs our era's software disaster woes.

As the technical process improves, we mistakenly conclude that we have solved our problems. But the human process surrounding the technical process remains frustrating and intractable. If, as Frederick Brooks said, "the hard thing about building software is deciding what to say, not saying it," then it doesn't matter how closely you model programming on bridge building. The builders of bridges are little better than the builders of software at "deciding what to say." They are people, too.

Freeman Dyson, the scientist, on stage for a talk at an open source software conference in 2004, was asked whether the technological systems humankind creates will ever grow so complex that we can't understand them. "The machines will become more and more complicated," he replied, "but they won't be as complicated as the societies that run them."

Certainly, the programming profession can and will continue to improve its technical skills and tools. Programmers love to linger on this part of their turf; there is little danger they will stop refining algorithms, improving back-end code, and tweaking development platforms.

But in doing so, they are sharpening axes or shaving yaks. They are not dealing with the essence of the matter. Sooner or later any mechanism, if it is to accomplish any function in the human world, must come out of its shell and communicate with people. "Ultimately, information systems only give value when they touch human beings," Jaron Lanier says. And when they do touch human beings, the prospect of perfection dissolves.

Software's essential difficulty, then, is the toll that human free will and unpredictability exact on technological progress. Though the rise of computing has led many scientists to view the human organism through its lens—imagining that the functions of the brain, for instance, map to the structure of a computer—individual people remain largely unprogrammable. They do things that the most imaginative programmers do not expect. And they want things that those programmers cannot anticipate.

■ ■ ■

In the summer of 2005, Bill Gates, talking at a Microsoft seminar in Singapore, described the research taking place at his company's labs in "computer implants." The first applications for such prosthetic electronics involve products that help blind and deaf people.

Doubtless there is good that can come from such research. Building reliable digital devices for medical applications is certainly within our power. But the broader prospect of implanting general-purpose computers in our bodies immediately raises questions and hackles. Whom would we trust to write the software? Whom would we trust to modify it? Will we be committing ourselves to a lifetime of upgrades and bug fixes?

Gates, the embodiment of software's conquest of modern life, seems to share these qualms. "One of the guys that works at Microsoft . . . always says to me, 'I'm ready. Plug me in,'" he said at the Singapore event. "I don't feel quite the same way. I'm happy to have the computer over there and I'm over here."

■ ■ ■

My three years on software time left me with no doubt that, for all the promise of exciting new methodologies and rigorous disciplines and break-through ideas, making software is still, and truly, hard. But I can hear the objection: There are different kinds of hard. Raising a child is hard. Burying a parent is hard. Birth and growing; living with other people or without them; trying to love them; failing to love them; accepting someone else's death; accepting your own—these things are hard. Software? That's a different universe of hard, a lesser one.

To which the only sane response is, yes. In fact, for those who enjoy the chance to tinker with towering puzzle palaces of code, the kind of difficulty software presents serves as a welcome contrast to the perplexing, unyielding frustrations of what geeks call our "wetware" lives. It is self-contained. It is rational. And it harbors no malice. In this sense, software isn't hard at all. It's a breeze, a "relief from the confusions of the world," as Ellen Ullman described it in her novel of programming, *The Bug*.

With its repetitious cycles and its infinite delays, the work of programming has always brought to mind the labor of Sisyphus endlessly rolling his boulder up the hill. It's the very image of futility. Yet the best-known modern interpretation of the Sisyphus myth, Albert Camus's existentialist take, tells us that the doomed Greek experiences a strange kind of joy in his work. "One must imagine Sisyphus happy," Camus says.

Most of the programmers I have interviewed and gotten to know in the course of my research are consistently, and sometimes unaccountably, optimistic about their work. If they are Sisyphuses, they are happy ones.

■ ■ ■

All things fall and are built again,
And those that build them again are gay.

—*William Butler Yeats, "Lapis Lazuli"*

Computers entered the popular imagination in the first decades of the era of computing, the 1950s and '60s, as marvels of infallibility. Error belonged to human beings; the computer did not make mistakes. Those humans who worked closely with computers always understood that the truth was far more complex: that the regions of human error and of computer precision had no discernible borders on any conceivable map; that one bled into the other; and that computers, remarkable and unique in so many ways, did not and could not by themselves transcend the capabilities and limitations of their creators.

In more recent decades we have all begun to work closely with computers, and the myth of computer infallibility has lost its power. Today, we have begun to grasp that the software which runs our world is no more perfectible than the hands and minds of the people who build and program it. Those people still have plenty more to learn about how to make that software better. But the goal of making it vastly, transformingly better—so much better that we can't even recognize it anymore; so much better that we could finally say about software for some particular job or activity, "It's done"—that goal looks unattainable.

Realizing this could overwhelm programmers with a sense of despair, but it doesn't. There is something liberating in the thought that every effort to make software must ultimately prove incomplete. It means that there is always room for an additional try, always an opportunity for a new generation to take another bite at the problem, always someone somewhere who will welcome "Yet Another" program.

In the spring of 2002, around the time Mitch Kapor and the early members of the Chandler team were beginning to zero in on their new software's architecture, Kapor made the tech news headlines for something entirely different: He entered into a Long Bet about the prospects for artificial intelligence. Long Bets were a project of the Long Now Foundation, a nonprofit organization started by Whole Earth Catalog creator Stewart Brand and a group of digital-age notables as a way to spur discussion and creative ideas about long-term issues and problems. As the project's first big-splash Long Bet, Kapor wagered $20,000 (all winnings earmarked for worthy nonprofit institutions) that by 2029 no computer or "machine intelligence" will have passed the Turing Test. (To pass a Turing Test, typically conducted via the equivalent of instant messaging, a computer program must essentially fool human beings into believing that they are conversing with a person rather than a machine.)

Taking the other side of the bet was Ray Kurzweil, a prolific inventor responsible for breakthroughs in electronic musical instruments and speech recognition who had more recently become a vigorous promoter of an aggressive species of futurism. Kurzweil's belief in a machine that could ace the Turing Test was one part of his larger creed—that human history was about to be kicked into overdrive by the exponential acceleration of Moore's Law and a host of other similar skyward-climbing curves. As the repeated doublings of computational power, storage capacity, and network speed start to work their magic, and the price of all that power continues to drop, according to Kurzweil, we will reach a critical moment when we can technologically emulate the human brain, reverse-engineering our own organic processors in computer hardware and software. At the same time, biotechnology and its handmaiden, nanotechnology, will be increasing their powers at an equally explosive rate.

When these changes join forces, we will have arrived at a moment that Kurzweil, along with others who share his perspective, calls "the Singularity." The term, first popularized by the scientist and science fiction author Vernor Vinge, is borrowed from physics. Like a black hole or any similar rent in the warp and woof of space-time, a singularity is a disruption of continuity, a break with the past. It is a point at which everything changes, and a point beyond which we can't see.

Kurzweil predicts that artificial intelligence will induce a singularity in human history. When it rolls out, sometime in the late 2020s, an artificial intelligence's passing of the Turing Test will be a mere footnote to this singularity's impact—which will be, he says, to generate a "radical transformation of the reality of human experience" by the 2040s.

Utopian? Not really. Kurzweil is careful to lay out the downsides of his vision. Apocalpytic? Who knows—the Singularity's consequences are, by definition, inconceivable to us pre-Singularitarians. Big? You bet.

It's easy to make fun of the wackier dimension of Kurzweil's digital eschatology. His personal program of life extension via a diet of 220 pills per day—to pickle his fifty-something wetware until post-Singularity medical breakthroughs open the door to full immortality—sounds more like something out of a late-night commercial pitch than a serious scientist's choice. Yet Kurzweil's record of technological future-gazing has so far proven reliable; his voice is a serious one. And when he argues that "in the short term we always underestimate how hard things are, but in the long term we underestimate how big changes are," he has history on his side.

But Kapor thinks Kurzweil is dead wrong. He thinks the whole project of "strong artificial intelligence," the effort from the 1950s to the present to replicate human intelligence in silicon and code, remains a folly. In an essay that explained his side of the Long Bet wager, Kapor wrote that the entire enterprise has misapprehended and underestimated human intelligence.

As humans:

- ▸ We are embodied creatures; our physicality grounds us and defines our existence in a myriad of ways.
- ▸ We are all intimately connected to and with the environment around us; perception of and inter-action with the environment is the equal partner of cognition in shaping experience.
- ▸ Emotion is as or more basic than cognition; feelings, gross and subtle, bound and shape the envelope of what is thinkable.
- ▸ We are conscious beings, capable of reflection and self-awareness; the realm of the spiritual or trans-personal (to pick a less loaded word) is something we can be part of and which is part of us.

How, Kapor wrote, could any computer possibly impersonate a human being whose existence is spread across so many dimensions? Or fool a human judge able to "probe its ability to communicate about the quintes-sentially human"?

"In the end," he declared, "I think Ray is smarter and more capable than any machine is going to be."

■ ▌ ▐

Kapor's arguments jibe with common sense on a gut level (that very phrase embodies his point). It will be another quarter century before we know if they are right.

By then I should finally be traveling over a new Bay Bridge. Chandler may have become a thriving open source ecosystem or may have been interred in the dead-code graveyard. And maybe, just maybe, the brains and talents and creativity of the world's programmers will have found a way out of the quicksand hold of software time.

But I'm not making any bets.

In 2005, three years after placing his Long Bet with Kurzweil, years spent clambering through the trenches of real-world software development, Mitch Kapor stands by his wager.

"I would double down with Ray in an instant," he says. "I don't run into anybody who takes his side. Now, I don't talk to people inside the Singularity bubble. But just your average software practitioner, whether they're twenty-five or forty-five or sixty-five—nobody takes his side of the bet."

A half-smiling glint—of innocent mischief, and also perhaps of hard-earned experience—widens his eyes. "The more you know about software, the less you would do that."

POSTSCRIPT TO THE PAPERBACK EDITION

I n early 2006, Mitch Kapor and the OSAF management group reviewed their work on Chandler 0.6 and concluded that their program remained too much of a "science project." For all the progress they'd made, it hadn't really proved edible as dogfood. Kapor and his assistant kept using it for brief periods until they hit some problem that stopped them. "It's like I was the test pilot," Kapor said. "Someone's got to get up and try to fly the new airplane."

The new plan: OSAF wouldn't ship another major release of Chandler until it could say, confidently, that the program was ready for the real world. Instead of calling the next release 0.7, it would be named "Preview," and would be the equivalent of a beta version—the last big release before 1.0. Preview would be targeted to the needs of small work groups, and all potential features would be ruthlessly evaluated by that criterion. OSAF would keep plugging away at Chandler's desktop program, the newer Cosmo server, and the Web-based interface—and not declare Preview "done" until it finally delivered a calendar that worked well enough for adventurous early adopters outside OSAF to begin using.

That ended up taking a full twenty-one months. Outside OSAF, only a handful of people kept up with the interim developers' releases of Chandler. In June 2006, one outsider, Hank Williams, posted an irate message to one of the Chandler mailing lists about the program's poor performance on his brand-new Mac: "Chandler is the slowest piece of software I have ever seen. It is truly frightening. . . . I can't understand why any app in 2006 could be so slow. I am concerned that years after this effort began, you 'can't get there from here.'"

Ted Leung responded, "Those of us using the system on a daily basis are *painfully aware* of how far we have to go." Katie Parlante patiently explained to Williams that his new Mac was one of Apple's latest—built, unlike previous Macs, around Intel processors—and that OSAF hadn't yet optimized Chandler for the new machines.

Rewriting Chandler to suit the new Mac was one example of the sort of unplanned but unavoidable work that kept piling up around the project. Another set of unpredictable problems arose around coordinating the Chandler desktop application with the server. Making Chandler usable required both accelerating its performance with countless code tweaks and improving its interface with countless design tweaks. And, of course, there were myriad bugs to be fixed, or at least carefully triaged.

And so Preview, originally intended for the end of 2006, kept slipping. The team recalibrated its schedule to aim first for a March 2007 release, then for June. Preview finally shipped in September, as I'm writing this postscript.

The new Chandler was, finally, a stable, functioning calendar program with some unusual features. It worked with the Chandler Hub—a model service using the Cosmo server (now rechristened as the Chandler Server) and operated by OSAF itself for Chandler's users—to synchronize multiple copies of the Chandler desktop application. The Hub also synced your desktop Chandler with the new Web-based interface, so that you could view and edit your Chandler data from any computer with an Internet connection.

The achievement was late but real. And beyond the mundane calendar functionality, users had a much-improved dashboard to fool around with, offering a glimpse of what remained of Chandler's original, grand vision. The dashboard now clearly grouped all your items—calendar events, notes, and tasks—into "Now," "Later," and "Done" categories. Stamping, or transmuting one kind of item into another, also worked pretty well.

But the dashboard, and indeed much of the Chandler interface, was tough for new users to grasp. It had its own language of icons—one that was innovative, but neither familiar nor intuitive. OSAF's developers also knew that, in other ways, Preview was simply an incomplete product. For instance, it lacked a month view—you could only view your calendar by day or by week—and you couldn't yet print from it. Chandler included rudimentary email functions to assist users who wanted to start sharing calendars, but it was in no shape to serve as anyone's primary email tool.

Still, it was a program that the everyday nonprogrammer could recognize as a working piece of software. As Ted Leung wrote on his blog: "Over the years, many people have said to me, 'Let me know when Chandler is usable.' This is your notice that we now consider Chandler usable."

The question no one at OSAF could answer after all this time and effort was, *how* usable? How would Chandler be received by the open source programmers it hoped to inspire, and the early-adopter users it aimed to serve? What kind of community, if any, would develop around it?

At least now, with Preview out the door, answers to these questions would emerge, as Kapor put it, "empirically, rather than theoretically." Meanwhile, the OSAF team started grappling with the question of its own future—an existential challenge that was routine for most software projects but that OSAF had been shielded from throughout its life. At the start of the Preview cycle, Kapor set a limit on his funding for OSAF: he would commit enough money to carry OSAF, at its current spending rate, through roughly the end of 2008. "A healthy project is one that finds its own means of support," he said. "I don't believe in the long-term patronage model."

What sort of future did Chandler have without Kapor paying for an extensive (and expensive) team? As long as their heads were down in the struggle to finish Preview, the developers could set that concern aside. But now a candid and sometimes heated debate broke out on the mailing list. If time was limited, which was more important to Chandler's future: devising a sound software architecture to make the project easy for programmers to work on, at the risk of a delay in getting working software into users' hands? Or delivering such a program as quickly as possible to users, at the risk of producing unmanageable code that developers wouldn't want to maintain or improve? The only sane course was some balance between the two. But what balance, exactly?

In a lengthy message to the developers' list, Phillip Eby argued that Chandler's code had so many problems that the team had to choose between "delivering a certain feature set, after which time the project will no longer effectively grow or be maintained," or undertaking an ambitious revamp in order to make the code base more approachable for the volunteer developers its future would eventually depend upon.

Eby declared that he was simply mapping out tradeoffs. But the choices he framed led ineluctably to the conclusion that it was time to rip out Chandler's foundations again.

Andi Vajda responded with a rant satirizing OSAF's "long tradition of being harsh on ourselves": "Yes, our architecture sucks, yes our UI sucks, yes our implementation sucks, yes our features suck, yes our schedule sucks, yes our product sucks, yes we all suck, yes software development sucks, yes the world sucks. Now what? . . . Would we do everything the same way were we to start over? I sure don't think so. The last thing we need though, right now, is another big bang, another 'let's start from scratch with a new approach' project." Vajda argued for more frequent and regular releases and a "users-first" approach to prioritizing work.

Parlante cooled things down via private email and let the team know that Chandler's future depended on building a core of "happy users."

"There's this common thing that happens," Parlante explained, "where developers just feel burnt out with the code base as it is, with all the things that they've been working against. It's sort of like, my baby's ugly! But we need to be focused on the first 100 users, and turning them into 1,000 users, and then 10,000 users. And the focus needs to be on fixing any problems that get in the way of that."

About five years had passed since the public unveiling of the ambitious Chandler project. The Bugzilla bug tracker, which had counted 4,732 bugs as I wrapped up the hardcover edition of this book, was nearing 11,000. (Of these, only 1,642 were marked "open." And OSAF had settled on using Bugzilla to track tasks as well as bugs, so that accounts for a significant chunk of the list.)

I have played with each new release of Chandler but have not adopted it for my own use. For now, Google Calendar does the job for me. It has plenty of drawbacks, but it passes the "good enough" test. As I begin work on my next book, I'm still organizing my research with the discontinued Ecco Pro, and that old program has, improbably, begun a new, entropy-defying metamorphosis.

In April, a programmer who goes by the handle "slangmgh" posted a brief note to a mailing list of Ecco devotees: "I write little utility, have upload to the files directory! It's only work for EccoPro v4.01." (Slangmgh was plainly not a native English speaker; he provided no return address, just a link to a profile page on Yahoo in Chinese.)

The "little utility" was actually a significant new program named Ecco Extension, or EccoExt. It fixed some of the problems that had nagged users in the decade since commercial work on Ecco had ceased, and added some useful new functions on top of the old program. Ecco Pro was not an open source program, but it provided enough "hooks," or means for a programmer to access its data directly, to make EccoExt possible.

Over the following weeks and months, slangmgh worked doggedly on his project, sometimes posting new versions of his code each night. He found a way to embed in Ecco an open source scripting language called Lua, thereby opening new ways to further extend the program and automate routine tasks. Another contributor who called himself "yoursowelcomethanks" began collaborating; he peppered his messages with bursts of enthusiastic capitalization and exclamation marks, lauding EccoExt's "super powers." The small community of die-hard Ecco users on the mailing list got used to reporting bugs and problems in the latest builds of EccoExt. Occasionally some new feature slangmgh had added was too hard to figure out, so the mailing-list regulars started writing their own explanations.

I found myself enthralled by the Ecco resurrection story. Maybe it was the programmers' odd combination of obsession and marginality—they could have walked in from a William Gibson novel. Maybe it was the speed and casual nature of EccoExt's process—such a contrast to Chandler's sobering story. Or maybe it was that EccoExt's steady stream of small improvements was making my work easier in a palpable way.

EccoExt wasn't much, really—mostly the work of a single developer, laboring over one obscure program, for one small mailing list of users. But for a brief spell, I almost forgot that software was hard.

NOTES

All quotations from Open Source Applications Foundation staff and descriptions of scenes at OSAF in this book are drawn from personal observation or based on personal interviews. An identical version of these notes, with active hyperlinks, is located at http://www.dreamingincode.com.

About the subtitle: "Two dozen programmers" is a rough tally of the size of the Chandler development team, which started out much smaller and fluctuated over time. "Three years" represents the time I spent observing the Chandler project from January 2003 through December 2005. "4,732 bugs" is the number of bugs entered into the Chandler Bugzilla database on the date I completed writing the manuscript for this book; the number has since climbed.

CHAPTER 0
SOFTWARE TIME

1   The game Sumer (also known as Hamurabi or Hammurabi) is documented in Wikipedia at http://en.wikipedia.org/wiki/Hamurabi. Full Basic code for the game can be found in David H. Ahl, ed., *BASIC Computer Games* (Creative Computing, 1978).

3   Salon's content management software is documented in an article in the online magazine Design Interact at http://www.designinteract.com/features_d/salon/index.html. Chad Dickerson wrote about it in his InfoWorld blog at http://weblog.infoworld.com/dickerson/000170.html.

4   On "flow," see Mihaly Csikszentmihalyi, *Flow: The Psychology of Optimal Experience* (Harper Perennial, 1991).

5   The Association for Computing Machinery's Hello World page is at http://www2.latech.edu/~acm/HelloWorld.shtml.

6   Knuth's "Software is hard" appears in a number of versions of his "Theory and Practice" talk, for example on p. 134 of Donald E. Knuth, *Selected Papers on Computer Science* (CSLI Publications/Cambridge University Press, 1996). The explanation of why programmers count from zero is from a Web page titled So You've Hired a Hacker by Jonathan Hayward of Fordham University, at http://jonathanscorner.com/writings/hacker/hacker4.html. A. P. Lawrence offers a more technical explanation at http://aplawrence.com/Basics/count fromzero.html.

6   "Maybe you noticed that I've called this Chapter 0": The occasional practice among programmers of starting books with a Chapter 0 appears to have originated with the classic programming text *The C Programming Language*, by Brian W. Kernighan and Dennis M. Ritchie (Prentice-Hall, 1978). Ellen Ullman also used it in her essay collection *Close to the Machine: Technophilia and Its Discontents* (City Lights, 1997).

8   Details on the construction of the new Bay Bridge are from http://www.newbaybridge.org/.

9   The National Institute of Standards and Technology study is at http://www.nist.gov/public_affairs/releases/n02-10.htm.

9   "Our civilization runs on software" is widely attributed to Bjarne Stroustrup. One original source is in slides from a course he teaches at Texas A&M University where he is a professor: http://courses.cs.tamu.edu/petep/1_programming _06a.pdf.

9   The Maurice Wilkes quote is from p. 145 of M. V. Wilkes, *Memoirs of a Computer Pioneer* (MIT Press, 1985) as cited in M. Campbell-Kelly, "The Airy Tape: An Early Chapter on the History of Debugging," *Annals of the History of Computing* 14: 4 (1992), pp. 18–28. That article is available at http://www.dcs.warwick.ac.uk/~mck/Personal/CampbellKelly1992.pdf.

10   Frederick P. Brooks, "No Silver Bullet: Essence and Accidents of Software Engineering," *Computer,* 20: 4 (April 1987), pp. 10–19. Also reprinted in *The Mythical Man-Month Anniversary Edition* (Addison Wesley, 1995). Available online at http://www-inst.eecs.berkeley.edu/~maratb/readings/NoSilverBullet.html.

CHAPTER 1
DOOMED

14   Netscape developers as a legion of the doomed: See Jamie Zawinski, Netscape Dorm, at http://www.jwz.org/gruntle/nscpdorm.html. The conference OSAF staffers attended was the O'Reilly Open Source Conference held in July 2003.

17   Brooks's Law can be found on p. 25 of Frederick Brooks, *The Mythical Man-Month Anniversary Edition* (Addison Wesley, 1995). "The very unit of effort . . . deceptive myth" is on p. 16.

17   "Men and months are interchangeable": Brooks, p. 16.

18   "Regenerative scheduling disaster": Ibid., p. 21.

18   "Therein lies madness": Ibid., p. 25.

18   "The bearing of a child takes . . .": Ibid., p. 17.

18   "conceptual integrity": Ibid., p. 42.

21   Bill Gates's comments on the GPL as Pac-Man were widely reported in 2001, for instance on CNET News.com at http://news.com.com/2100-1001-268667.html.

21   Torvalds's "Just a hobby" quotation is from his 1991 message announcing the Linux project to the comp.os.minix newsgroup. It is archived many places online, e.g. at http://www.linux.org/people/linus_post.html.

22   "that purists call GNU-Linux": a good account of this issue is in Wikipedia at http://en.wikipedia.org/wiki/GNU/Linux_naming_controversy.

The "Free speech" vs. "Free beer" argument is outlined at http://www.gnu.org/ philosophy/free-sw.html.

23   All quotations from "The Cathedral and the Bazaar" may be found in the online version at http://www.catb.org/~esr/writings/cathedral-bazaar/cathedral-bazaar/index.html.

24   Apache market share is tracked by the Netcraft survey at http://news.netcraft.com/archives/web_server_survey.html.

25   "The total cost of maintaining": Brooks, p. 121.

26   "anyone may use it, fix it, and extend it": Ibid., p. 6.

26   "incompletely delivered": Ibid., p. 8.

30   Michael Toy's blog entry from June 26, 2003, is at http://blogs.osafoundation.org/blogotomy/2003_06.html.

31   "A baseball manager recognizes": Brooks, p. 155.

CHAPTER 2
THE SOUL OF AGENDA

33   "sell sugar water": Steve Jobs's pitch to John Sculley has become the stuff of legend. The transcript from the PBS documentary *Triumph of the Nerds*, in which Sculley himself reports it, is a relatively primary source: http://www.pbs.org/nerds/part3.html.

35   Kapor's estimated $100 million: *Business Week*, May 30, 1988, p. 92.

36   "It's important to understand": David Gans's interview with Kapor is at http://www.eff.org/Misc/Publications/John_Perry_Barlow/HTML/barlow_and_kapor_in_wired_interview.html.

36   "extricating myself from my own success": Kapor interview in *Inc.*, January 1, 1987.

37 Lotus Agenda: general background on the program is collected at http://
home.neo.rr.com/pim/alinks.htm. The program is still available via http://
www.bobnewell.net/nucleus/bnewell.php?itemid=186.

37 The principles behind Agenda are outlined in a development document
from the original team, available at http://home.neo.rr.com/pim/article1.htm.
James Fallows's article on Agenda appeared in the *Atlantic* in May 1992.

41 "In science the whole system builds": Linus Torvalds, quoted in *Business
Week*, August 18, 2004, at http://www.businessweek.com/technology/
content/aug2004/tc20040818_1593.htm.

42 Vannevar Bush's "As We May Think" first appeared in the *Atlantic* in July
1945. It is available at http://www.theatlantic.com/doc/194507/bush.

My account of Douglas Engelbart's work draws on readings from his work
collected at the Bootstrap Institute Web site at http://www.bootstrap.org/, as
well as the accounts in Thierry Bardini, *Bootstrapping* (Stanford University
Press, 2000); Howard Rheingold, *Tools for Thought* (Simon & Schuster, 1985);
and John Markoff, *What the Dormouse Said* (Viking, 2005).

43 The video of Engelbart's 1968 demo is at http://sloan.stanford.edu/mouse
site/1968Demo.html.

43 "store ideas, study them": From the Invisible Revolution Web site, devoted to
Engelbart's ideas, at http://www.invisiblerevolution.net/nls.html.

43 "successful achievements can be utilized": From the "Whom to Aug-
ment First?" section of Engelbart's 1962 paper, "Augmenting Human Intellect: A Conceptual Framework," at http://sloan.stanford.edu/mousesite/
EngelbartPapers/B5_F18_ConceptFrameworkPt4.html.

44 "an improving of the improvement process": Bootstrap Institute home page
at http://www.bootstrap.org/.

44 "the feeding back of positive research": In the "Basic Regenerative Feature"
section of Engelbart's 1962 paper, "Augmenting Human Intellect," at http://
sloan.stanford.edu/mousesite/EngelbartPapers/B5_F18_ConceptFramework
Pt4.html.

45 "Some astonished visitors": Bardini, *Bootstrapping*, p. 145.

46   "Engelbart, for better or worse": Alan Kay, quoted in Bardini, *Bootstrapping*, p. 215.

46   Jaron Lanier's story about Marvin Minsky is from a video of the "Engelbart's Unfinished Revolution" seminar at Stanford, 1998, available at http://unrev .stanford.edu/.

47   The FBI's software disasters were recounted in the January 14, 2005, *New York Times*, available at http://www.nytimes.com/2005/03/09/politics/09fbi.html? ex=1268110800&en=efa63369bfa14be8&ei=5090&partner=rssuserland. IEEE *Spectrum* analyzed the failure in its September 2005 issue, at http:// www.spectrum.ieee.org/sep05/1455.

48   The IRS's software troubles were chronicled in the December 11, 2003, *New York Times*, available at http://www.nytimes.com/2003/12/11/business/ 11irs.html?ex=1386478800&en=39d69ddfd8171e0c&ei=5007&partner= USERLAND. CIO *Magazine*'s detailed report from April 1, 2004, is at http://www.cio .com/archive/040104/irs.html?printversion=yes.

49   The U.K. pension system crash was widely reported in the United Kingdom, for instance in this *Guardian* article from November 26, 2004: http:// politics.guardian.co.uk/homeaffairs/story/0,11026,1360163,00.html.

49   The McDonald's Innovate disaster was explored in *Baseline*, July 2, 2003, at http://www.baselinemag.com/article2/0,3959,1191808,00.asp.

49   *Computerworld* covered the Ford Everest system's story on August 18, 2004, at http://www.computerworld.com/softwaretopics/erp/story/0,10801,95335, 00.html.

50   The Standish Group's CHAOS report data is collected at http://www .standishgroup.com/public.php, including the original 1994 report and several updates.

51   Robert L. Glass, *Software Runaways* (Prentice Hall, 1998). Edward Yourdon, *Death March* (Prentice Hall, 1997). Robert N. Britcher, *The Limits of Software* (Addison Wesley, 1999).

51   "may have been the greatest": Britcher, *The Limits of Software*, p. 163.

51   "like replacing the engine on a car": Ibid., p. 181.

51   "The software could not be written": Ibid., p. 185.

52   "you can read the *Iliad*": Ibid., p. 172.

52   "One engineer I know": Ibid., pp. 168–69.

53   "All programmers are optimists": Frederick Brooks, *The Mythical Man-Month Anniversary Edition* (Addison Wesley, 1995), p. 14.

54   Kapor delivered his "Software Design Manifesto" at the 1990 PC Forum conference. It was later published in Terry Winograd, *Bringing Design to Software* (Addison Wesley, 1996). It can be found at http://hci.stanford.edu/bds/1-kapor.html.

55   "We took the plan out": From "Painful Birth: Creating New Software Was Agonizing Task for Mitch Kapor Firm" by Paul B. Carroll, *Wall Street Journal*, May 11, 1990.

CHAPTER 3

PROTOTYPES AND PYTHON

56   "a crew of twenty people": Artist Chris Cobb's project at the Adobe Bookstore in San Francisco is chronicled at the McSweeney's Web site at http://www.mcsweeneys.net/links/events/chriscobb.html.

59   Information about the Semantic Web and RDF is at http://www.w3.org/2001/sw/.

61   "plan to throw one away" and "promise to deliver a throwaway": Frederick Brooks, *The Mythical Man-Month* (Addison Wesley, 1995), pp. 115–16.

64   "The programmer, like the poet": Ibid., p. 7.

64   "The lunatic, the lover, and the poet": William Shakespeare, *A Midsummer Night's Dream*, Act V, sc. i.

66   "The process of combining multiple": The phrase is from Wikipedia's definition of "Abstraction (computer science)." At the time of writing this phrase no longer appears on that page, but it may be found here: http://en.wikipedia

.org/w/index.php?title=Abstraction_%28computer_science%29&oldid=12117920.

66   "This is what programmers do": Eric Sink, "The .Net Abstraction Pile," blog entry at http://biztech.ericsink.com/Abstraction_Pile.html.

67   "close to the machine": Title of Ellen Ullman's book *Close to the Machine* (City Lights, 1997).

67   "virtually eliminate coding and debugging": The words are from page 2 of a 1954 report titled "Preliminary Report, Specifications for the IBM Mathematical FORmula TRANslating System, FORTRAN," as cited later in John Backus, "The History of Fortran I, II, and III," *IEEE Annals of the History of Computing* 20: 4 (Oct.–Dec., 1998) pp. 68–78, available at http://doi.ieee computersociety.org/10.1109/85.728232.

67   "hand to hand combat with the machine": Backus's phrase is quoted in Steve Lohr, *Go To: The Story of the Math Majors, Bridge Players, Engineers, Chess Wizards, Maverick Scientists and Iconoclasts—the Programmers Who Created the Software Revolution* (Basic, 2001), p. 13.

69   Moore's Law is outlined at Intel's Web site: http://www.intel.com/technology/silicon/mooreslaw/.

72   Quotations from Guido van Rossum are from a talk given February 17, 2005, at the Software Development Forum in Palo Alto, California. Audio is available at http://www.itconversations.com/shows/detail545.html.

72   Larry Wall's talk was at the O'Reilly Open Source Conference, Portland, Oregon, July 2004. Text is available at http://www.perl.com/pub/a/2004/08/18/onion.html.

73   one observer's characterization: The observer is Danny O'Brien in his NTK newsletter from August 6, 2004, at http://www.ntk.net/2004/08/06/.

74   "I spent a few weeks trying": Benjamin Pierce in a June 2001 message on a private mailing list; full quote confirmed in email to author.

77   "Guido's time machine": Eric Raymond's Jargon File defines it at http://www.catb.org/jargon/html/G/Guido.html.

79 "When you program, you spend": Paul Graham, "The Python Paradox," August 2004, at http://www.paulgraham.com/pypar.html.

79 Vaporware Hall of Fame: Jon Zilber in *MacUser*, January 1, 1990.

81 Dan Gillmor's piece: "Software Idea May Be Just Crazy Enough to Work," *San Jose Mercury News*, October 20, 2002.

81 The original Slashdot posting and discussion is at http://slashdot.org/ articles/02/10/20/1827210.shtml?tid=99. Complete archives of OSAF's mailing lists can be accessed at http://www.osafoundation.org/mailing_lists.htm.

83 "We will first put out code": Kapor's blog post from November 14, 2002, is at http://blogs.osafoundation.org/mitch/000044.html#000044.

CHAPTER 4
LEGO LAND

87 David/Rys McCusker's blog postings are no longer online.

93 "In the future, programs will be built": James Noble, "The Lego Hypothesis," slides from talk at JAOO Conference, September 2004, at http://www .jaoo.org/jaoo2004/speakers/show_speaker.jsp?oid=58.

The account of James Noble and Robert Biddle's research disproving the Lego hypothesis is drawn from James Noble, Robert Biddle, Alex Potanin, and Marcus Frean, "Scale-free Geometry in OO Programs," *Communications of the ACM*, May 2005, available at http://portal.acm.org/citation.cfm?id=1060716 and also http://www.mcs.vuw.ac.nz/~marcus/manuscripts/CACM.pdf.

95 "in order that every user": Maurice Wilkes, *Memoirs*, as cited in James Noble and Robert Biddle, "No Name: Just Notes on Software Reuse," at http://www .mcs.vuw.ac.nz/comp/Publications/CS-TR-03-11.abs.html.

95 Robert Glass's observations on "reuse-in-the-small" are from Robert L. Glass, *Facts and Fallacies of Software Engineering* (Addison Wesley, 2003), pp. 43–49.

95 "It is not a will problem": Ibid., p. 47.

95 "transform programming from a solitary": This and later quotes from Brad Cox are from Cox's "Planning the Software Industrial Revolution," *IEEE*

Software, November 1990, and also at http://virtualschool.edu/cox/pub/PSIR/.

96 Brad Cox, *Superdistribution: Objects as Property on the Electronic Frontier* (Addison Wesley, 1995).

97 "They do have an economic model": Author interview with Brad Cox, June 2005.

97 "Unfortunately, most programmers like to program": Larry L. Constantine, *Constantine on Peopleware* (Prentice Hall, 1995), pp. 123–24.

99 "Keeping up with what's available": Ward Cunningham, quoted by Jon Udell in his *InfoWorld* blog at http://weblog.infoworld.com/udell/2004/05/21.html#a1006.

101 "People keep pretending they can": These lines by Ted Nelson are widely distributed on the Net, and the word *intertwingle* appears frequently in Nelson's writing, but the original source of the full quotation is obscure. One source cited is p. 45 of the first (1974) edition of his book *Computer Lib/Dream Machines*. There are two discussions of the quote's origins at http://www.bootstrap.org/dkr/discussion/3260.html and http://listserv.linguistlist.org/cgi-bin/wa?A2=ind0204b&L=ads-l&D=0&P=5140.

107 "I've been referring to Chandler": Kapor's blog posting from March 29, 2003, is at http://blogs.osafoundation.org/mitch/000139.html#000139.

CHAPTER 5
MANAGING DOGS AND GEEKS

117 "They get along well with other dogs": The labradoodle description is from http://www.dogbreedinfo.com/labradoodle.htm.

120 "We are already too large": From Mitch Kapor's blog posting, Making Design Decisions: Some Principles, December 29, 2002, at http://blogs.osafoundation.org/mitch/000097.html#000097.

123 "The name for my pain": Rick Kitts, "I Know Why Software Sucks," February 7, 2004, at http://www.artima.com/weblogs/viewpost.jsp?thread=32878.

125 The scarecrow image is no longer at Michael Toy's home page, but it can be found at http://toyblog.typepad.com/about.html.

125 "The transition from programmer to manager": This is part of the title of Toy's Blogotomy blog at http://blogs.osafoundation.org/blogotomy/.

125 "Michael is the last person": This is at http://www.undignified.org/people.html.

126 "the ground on which we are": Michael Toy blog posting, June 26, 2003, at http://blogs.osafoundation.org/blogotomy/000248.html.

126 "Management is about human beings": Peter Drucker, "Management as Social Function and Liberal Art," in The Essential Drucker (Harper Business, 2001), p. 10.

127 Bill Atkinson's "-2000" lines of code: From Andy Hertzfeld, Revolution in the Valley: The Insanely Great Story of How the Mac Was Made (O'Reilly, 2005), p. 65 . Also at http://www.folklore.org/StoryView.py?project=Macintosh&story=Negative_2000_Lines_Of_Code.txt.

128 "management by wandering around": Tom Peters in a blog posting from September 6, 2005, at http://www.tompeters.com/entries.php?note=008106.php.

129 "With manufacturing, armies, and traditional hardware": Watts Humphrey, "Why Big Software Projects Fail," CrossTalk, March 2005, at http://www.stsc.hill.af.mil/CrossTalk/2005/03/0503Humphrey.html.

130 According to wordorigins.org, geek "is a variant of geck, a term of Low German/Dutch origin that dates in English to 1511. It means a fool, simpleton, or dupe." Geck appears in Twelfth Night, Act V, scene i; a variant, geeke, turns up in Cymbeline, Act V, scene iv.

130 "one who eats (computer) bugs": The original definition of computer geek is from Eric Raymond, ed., The New Hacker's Dictionary, 3rd ed. (MIT Press, 1996), p. 120.

131 "Geek: A person who has chosen": The current definition of geek from the online Jargon File (source of the Hacker's Dictionary) is at http://www.catb.org/jargon/html/G/geek.html.

131 "To geek out on something": From Neal Stephenson, "Turn On, Tune In, Veg Out," *New York Times*, June 17, 2005, and also at http://www.nytimes.com/2005/06/17/opinion/17stephenson.html?ex=1276660800&en=a693ccc4ec 008424&ei=5090&partner=rssuserland&emc=rss.

131 "Some deranged fold of my lizard brain": Merlin Mann on his 43 Folders blog, September 15, 2004, at http://www.43folders.com/2004/09/15/home-comforts-illustrated-housekeeping-pr0n/.

132 Jennifer Tucker, Abby Mackness, Hile Rutledge, "The Human Dynamics of Information Technology Teams," in Crosstalk, February 2004, at http://www.stsc.hill.af.mil/crossTalk/2004/02/0402Tucker.html.

133 "A lot of people feel that": Abby Mackness presentation at the Systems & Software Technology Conference, Salt Lake City, April 2004.

133 Steve Silberman, "The Geek Syndrome," *Wired*, December 2001, at http://www.wired.com/wired/archive/9.12/aspergers_pr.html.

135 "The typical behavior of a student": Gerald Weinberg, *The Psychology of Computer Programming, Silver Anniversary Edition* (Dorset House, 1998), p. 50.

138 "I figured, OK, I'm running this repository": Ward Cunningham's talk at the OOPSLA Conference, October 2004, Vancouver, B.C.

139 The Portland Pattern Repository is at http://c2.com/ppr/.

140 The OSAF wiki is at http://wiki.osafoundation.org.

142 Brooks's discussion of the Tower of Babel is in *The Mythical Man-Month Anniversary Edition* (Addison Wesley, 1995), p. 74.

143 "There are a couple of dark sides": James Gosling, "Sharpen the Axe: The Dark Side," blog entry from January 4, 2005, at http://today.java.net/jag/page13.html#106.

143 The Jargon File entry on yak shaving is at http://www.faqs.org/docs/jargon/Y/yak-shaving.html.

CHAPTER 6

GETTING DESIGN DONE

148 The *Economist* wrote at length about the "mom test" on October 28, 2004, at http://www.economist.com/displaystory.cfm?story_id=3307430.

149 "No one is speaking for the poor user": From Mitchell Kapor's "Software Design Manifesto" in Terry Winograd, *Bringing Design to Software* (Addison Wesley, 1996), and at http://hci.stanford.edu/bds/1-kapor.html.

153 For more on the origins of the Mozilla Foundation, see Wikipedia's entry at http://en.wikipedia.org/wiki/Mozilla_Foundation, and CNET coverage from January 13, 2005, at http://news.com.com/2102-7344_3-5519612 .html.

158 Two useful documents on the early history of CPIA are the original September 2003 design document at http://wiki.osafoundation.org/bin/view/ Journal/ChandlerPresentationAndInteractionArchitectureSep2003, and later documentation during the Chandler 0.3 cycle at http://wiki.osafoundation .org/bin/view/Projects/CpiaZeroPointThreeStatus.

159 "A favorite story at management meetings": Peter Drucker, *The Essential Drucker* (Harper Business, 2001), p. 113.

162 "Chandler 0.2 is about to ship": Michael Toy blog posting, September 19, 2003, at http://blogs.osafoundation.org/blogotomy/000388.html.

165 David Allen, *Getting Things Done* (Viking Penguin, 2001).

166 Information on David Gelernter's Lifestreams project is at http://www.cs .yale.edu/homes/freeman/lifestreams.html.

168 Andy Hertzfeld's Folklore site is at http://www.folklore.org.

169 "Moving your files from this machine": Bill Gates, quoted in *USA Today*, June 29, 2003, at http://www.usatoday.com/tech/news/2003-06-29-gates-longhorn _x.htm.

170 "Our head count has been fairly flat": Mitch Kapor blog posting on August 3, 2003, at http://blogs.osafoundation.org/mitch/000313.html#000313.

174 "Do you have any advice for people": Linus Torvalds, quoted in *Linux Times*,

June 2004. *Linux Times* has ceased publication. The article used to be at http://www.linuxtimes.net/modules.php?name=News&file=article&sid=145 and can be found via the Internet Archive's Wayback Machine at http://web.archive.org/web/20041106193140/http://www.linuxtimes.net/modules.php?name=News&file=article&sid=145.

CHAPTER 7
DETAIL VIEW

184 Simple things should be simple: This quotation is widely attributed to Alan Kay. I have been unable to trace its original source. It is also occasionally attributed to Larry Wall.

184 Clay Shirky wrote about Christopher Alexander's "A City Is Not a Tree" in the Many to Many blog on April 26, 2004, at http://many.corante.com/archives/2004/04/26/a_city_is_not_a_tree.php.

Alexander's article was originally published in *Architectural Forum*, April–May 1965. It is available online at http://www.arquitetura.ufmg.br/rcesar/alex/_cityindex.cfm.

188 The story of Donn Denman and the cancellation of MacBasic is at Folklore.org, at http://www.folklore.org/StoryView.py?project=Macintosh&story=MacBasic.txt.

196 Wikipedia defines *Foobar* at http://en.wikipedia.org/wiki/Foo_bar, and *fubar* at http://en.wikipedia.org/wiki/FUBAR.

196 "because people read these names": Ward Cunningham's talk at the OOPSLA Conference, October 2004, Vancouver, B.C.

197 Alec Flett first posted his parody of Hungarian notation on a Mozilla newsgroup in 1999. He repeated it in a blog posting from June 14, 2004, at http://www.flett.org/archives/2004/06/14/16.34.17/index.html.

198 Joel Spolsky traced the forking of Hungarian notation in "Making Wrong Code Look Wrong," May 11, 2005, at http://www.joelonsoftware.com/articles/Wrong.html. Charles Simonyi blogged his thinking on Hungarian

notation on July 8, 2005, at http://blog.intentionalsoftware.com/intentional
_software/2005/07/hungarian_notat.html.

204   "born with a silver spoon in its mouth": Esther Dyson's comment about On
Technology was in "On Technology Gets on Track," *Forbes*, June, 1991.

CHAPTER 8
STICKIES ON A WHITEBOARD

208   "Dogfooding . . . is something we do": This description of dogfooding is from
a February 21, 2003, blog posting by Scott Guthrie, a general manager in the
Microsoft Developer Division, at http://weblogs.asp.net/scottgu/archive/
2003/02/21/2743.aspx.

209   "We use what we build": Robert Taylor, quoted in Thierry Bardini, *Bootstrapping* (Stanford University Press, 2000), p. 154.

211   "allows users to collaboratively edit": Description from the main WebDAV
site at http://www.webdav.org/.

219   "hazmat > you know what would be really helpful": From the Chandler IRC
chat channel, July 14, 2004, at http://wiki.osafoundation.org/script/getIrc
Transcript.cgi?channel=chandler&date=20040714.

220   "People seem to be extraordinarily patient": Blog posting by Ted Leung from
July 29, 2004, at http://www.sauria.com/blog/2004/Jul/29.

224   There was extensive coverage of the "gutting" of Longhorn, including the
dropping of WinFS, in late August 2004. Microsoft Watch's article, titled
"Microsoft to Gut Longhorn to Make 2006 Delivery Date," is at http://www
.microsoft-watch.com/article2/0,1995,1640183,00.asp.

224   News.com interviewed Bill Gates at http://news.com.com/Gates%3A+
Longhorn+changed+to+make+deadlines/2008-1016_3-5327377.html.

224   A memo from Microsoft VP Jim Allchin on the changes was published
on Microsoft Watch at http://www.microsoft-watch.com/article2/0,1995,
1640602,00.asp.

CHAPTER 9
METHODS

239 Edsger Dijkstra, "Go To Statement Considered Harmful," *Communications of the ACM*, March 1968, at http://www.acm.org/classics/oct95/.

240 "necklace strung from individual pearls" and "As a slow-witted human being": Edsger Dijkstra, "Notes on Structured Programming," 1969, at http://www.cs.utexas.edu/users/EWD/ewd02xx/EWD249.PDF.

240 Edsger Dijkstra, "The Humble Programmer," 1972 Turing Award lecture, at http://www.cs.utexas.edu/users/EWD/ewd03xx/EWD340.PDF.

240 "each program layer is to be understood": Dijkstra, "Notes on Structured Programming."

241 "Unless developers plan and track": Watts Humphrey, "Why Big Software Projects Fail," CrossTalk, March 2005, at http://www.stsc.hill.af.mil/Cross Talk/2005/03/0503Humphrey.html.

241 "Most discussions of the knowledge worker's task": Peter Drucker, "The Effective Executive" (1966), in *The Essential Drucker* (Harper Business, 2001), p. 225.

242 Watts Humphrey's account of his IBM experience is from his "Reflections on a Software Life" in *In the Beginning: Recollections of Software Pioneers*, Robert L. Glass, ed. (IEEE Computer Society Press, 1998), p. 29 and ff.

243 "My daughter persuaded me to go": Humphrey interview in "Watts Humphrey: He Wrote the Book on Debugging," *Business Week*, May 9, 2005, at http://yahoo.businessweek.com/print/magazine/content/ 05_19/b3932038 _mz009.htm.

244 "An organization at Level 1": Humphrey's informal description is quoted in Mark Minasi, *The Software Conspiracy: Why Software Companies Put Out Faulty Products, How They Can Hurt You, and What You Can Do About It* (McGraw-Hill, 2000), pp. 48–49.

245 "The CMM reveres process, but ignores": James Bach, "The Immaturity of

CMM," *American Programmer,* September 1994, at http://www.satisfice.com/ articles/cmm.shtml.

245 "We all work for organizations": Slide set for a lecture by Watts Humphrey titled "Setting the Agile Context" at the XP Agile Universe Conference, Chicago, August 2002. Slides at http://www.xpuniverse.com/pdfs/SettingThe AgileContextWattsHumphrey.

246 "In all of modern technology": From a video distributed by the Software Engineering Institute, available at http://www.sei.cmu.edu/videos/watts/ DPWatts.mov.

246 Minasi, *The Software Conspiracy.*

247 The Mariner 1 bug is described at http://nssdc.gsfc.nasa.gov/nmc/tmp/ MARIN1.html.

248 James Gleick tells the story of the Ariane 5 bug at http://www.around.com/ ariane.html.

248 The Therac-25 bug is detailed in a paper by Nancy Leveson and Clark S. Turner in *IEEE Computer,* July 1993, at http://courses.cs.vt.edu/~cs3604/lib/ Therac_25/Therac_1.html.

248 The 1991 Patriot missile bug is well documented, for instance at http:// www.cs.usyd.edu.au/~alum/patriot_bug.html.

249 Jon Ogg's talk was at the Systems & Software Technology Conference, Salt Lake City, April 2004.

249 "The amount of software the Department of Defense": Barry Boehm at the Systems & Software Technology Conference, 2004. For more on the software patterns movement, see Erich Gamma, Richard Helm, Ralph Johnson, and John Vlissides, *Design Patterns: Elements of Reusable Object-Oriented Software* (Addison Wesley, 1995).

250 "programmers are like carpenters": Brian Hayes, "The Post-OOP Paradigm," *American Scientist,* March–April 2003, at http://www.americanscientist.org/ template/AssetDetail/assetid/17307.

250 "These have the advantages": From Kent Beck and Ward Cunningham, "A

Laboratory for Teaching Object-Oriented Thinking," from the OOPSLA '89 Conference Proceedings, October 1989, New Orleans, at http://c2.com/doc/oopsla89/paper.html.

251 Barry Boehm, "A Spiral Model of Software Development and Enhancement," in ACM SIGSOFT Software Engineering Notes, August 1986.

252 "Part of the purpose of the workshop": Brian Marick's blog posting from March 17, 2004, is at http://www.testing.com/cgi-bin/blog/2004/03/17#march17.

252 The Agile Manifesto is at http://agilemanifesto.org/.

253 "We were taking all these practices": Ron Jeffries's quote is from Sam Williams, "Totally Awesome Software?" in *Salon*, May 29, 2002, at http://archive.salon.com/tech/feature/2002/05/29/extreme_programming/index.html.

254 "Always implement things": The YAGNI principle is defined at http://xp.c2.com/YouArentGonnaNeedIt.html.

254 "Deliver Crap Quickly": From Robert Lefkowitz's blog posting titled "Extreme Programming Refactored" from April 2004; it used to be at http://r0ml.blogs.com/fot/2004/04/extreme_program.html and is now offline. Available at http://web.archive.org/web/20040810155153/http://r0ml.blogs.com/fot/2004/04/extreme_program.html.

255 "A 2004 study by two Pennsylvania": Phillip A. Laplante and Colin J. Neill, " 'The Demise of the Waterfall Model Is Imminent' and Other Urban Myths," *ACM Queue*, February 2004, at http://www.acmqueue.com/modules.php?name=Content&pa=showpage&pid=110.

256 "If you have even the slightest bit": Joel Spolsky blog posting from November 14, 2003, at http://www.joelonsoftware.com/items/2003/11/14.html.

256 "Beware of Methodologies": Joel Spolsky, "Big Macs vs. The Naked Chef," posting from January 18, 2001, at http://www.joelonsoftware.com/articles/fog0000000024.html. Also collected in *Joel on Software* (Apress, 2004), p. 237.

257 The Joel Test: Joel Spolsky posting from August 9, 2000, at http://www.joelonsoftware.com/articles/fog0000000043.html, and also in *Joel on Software*, p. 17.

263 37 Signals presented more on its philosophy in a 2006 PDF book, *Getting Real*, at https://gettingreal.37signals.com/.

263 Google's software development methods are outlined in Quentin Hardy, "Google Thinks Small," *Forbes*, November 14, 2005, at http://www.forbes.com/global/2005/1114/054A_print.html.

264 Nicholas Carr, "IT Doesn't Matter," *Harvard Business Review*, May 2003, at http://www.nicholasgcarr.com/articles/matter.html.

CHAPTER 10
ENGINEERS AND ARTISTS

270 All quotes are from the 1968 NATO software engineering conference report: P. Naur and B. Randell, eds., "Software Engineering: Report of a Conference Sponsored by the NATO Science Committee," Garmisch, Germany, October 7–11, 1968 (Scientific Affairs Division, NATO, 1969). The 1969 conference report: B. Randell and J. N. Buxton, eds., "Software Engineering Techniques: Report of a Conference Sponsored by the NATO Science Committee," Rome, Italy, October 27–31, 1969 (Scientific Affairs Division, NATO, 1970). Both reports and background material are available at http://homepages.cs.ncl.ac.uk/brian.randell/NATO/.

271 The Texas dispute over programmers calling themselves engineers is chronicled in "An Engineer by Any Other Name: Legislature to Decide if Computer Programmers Can Legally Use the Title," by R. G. Ratcliffe, *Houston Chronicle*, March 29, 2003.

272 "fully accepted that the term": Brian Randell, "The 1968/69 NATO Software Engineering Reports," at http://homepages.cs.ncl.ac.uk/brian.randell/NATO/NATOReports/index.html.

272 "The phrase 'software engineering' was deliberately chosen": 1968 conference report, p. 8.

272 "Unlike the first conference": Randell, "The 1968/69 NATO Software Engineering Reports."

272 "Bore little resemblance": 1969 conference report, p. 8.

272 Tom Simpson's "Masterpiece Engineering" is at http://homepages.cs.ncl.ac.uk/ brian.randell/NATO/NATOReports/index.html#Appendix.

273 A definitive discussion of the art/science dichotomy can be found in the preface to Steven Johnson, *Interface Culture* (HarperEdge, 1997).

275 For the etymology of *engineering*, see http://en.wikipedia.org/wiki/Engineering #Etymology and Webster's New World Dictionary (Simon & Schuster, 1984), p. 463.

275 "A Software Engineer, a Hardware Engineer": This joke is found in many locations online; for example: http://www.eff.org/Net_culture/Folklore/ Humor/engineer.joke.

276 " 'Software engineering' is something": L. Peter Deutsch, January 1999 ACM Fellow profile, at http://www.acm.org/sigsoft/SEN/deutsch.html.

277 physicists "deal with the absolute foundations": Alan Kay, Turing Award lecture at OOPSLA Conference, October 2004. Video is available at http:// www.acm.org/talks/AlanKay/KayTuring.htm.

277 "Hopper believed that programming": From her official biography page at http://www.hopper.navy.mil/grace/grace.htm.

278 "Software as we know it" and "There are two meanings": Charles Simonyi at "Programmers at Work Reunion" event, March 16, 2004, quoted in Scott Rosenberg, "Why Software Still Stinks," Salon, March 19, 2004, at http://archive .salon.com/tech/col/rose/2004/03/19/programmers_at_work/index.html.

279 "It's something like an architect": Clare Tristram, "Everyone's a Programmer," *Technology Review*, November 2003, at http://www.techreview.com/read _article.aspx?id=13377&ch=infotech.

279 "Somebody once asked me": Charles Simonyi interview with David Berlind, March 22, 2005, at http://blogs.zdnet.com/BTL/index.php?p=1190.

282 "All non-trivial abstractions" and "Abstractions do not really": Joel Spolsky, "The Law of Leaky Abstractions," November 11, 2002, online at http://joel onsoftware.com/articles/LeakyAbstractions.html and also in *Joel on Software* (Apress, 2004), p. 197.

283 "A well-known scientist": Stephen Hawking, *A Brief History of Time* (Bantam, 1988), p. 1.

284 Dr. Seuss, *Yertle the Turtle and Other Stories* (Random House, 1958).

284 "If builders built houses": Quotation widely attributed to Gerald Weinberg and confirmed in email to author.

285 Description of Alan Kay's presentation is from author's observation at the O'Reilly Emerging Technology Conference (ETech), April, 2003. Lisa Rein recorded the event; see http://www.lisarein.com/alankay/tour.html.

286 "We just don't know how": Kay, ETech talk.

286 "If you look at software today": From "A Conversation with Alan Kay," *ACM Queue,* December 2004–January 2005, at http://acmqueue.com/modules .php?name=Content&pa=showpage&pid=273&page=2.

287 "You can make a doghouse": Alan Kay, "The Computer Revolution Hasn't Happened Yet," keynote address at OOPSLA 1997, on the 25th anniversary of Smalltalk. Video available at http://video.google.com/videoplay?docid =2950949730059754521.

287 "It's all air and glass": Kay, OOPSLA 1997 keynote.

288 "We don't have to build pyramids": Kay, OOPSLA 2004 Turing Award lecture.

288 "Someday, we're going to invent": Alan Kay group interview at ETech 2003.

288 "Something like a computer": Kay, OOPSLA 1997 keynote.

289 "a reaction against the 'Indo-European'" and "In computer terms, Smalltalk": Alan Kay, "The Early History of Smalltalk," ACM SIGPLAN 1993, at http:// portal.acm.org/citation.cfm?coll=GUIDE&dl=GUIDE&id=155364.

289 "I made up the term object-oriented": Kay, OOPSLA 1997 keynote.

290 "At PARC, our idea was": "A Conversation with Alan Kay," *ACM Queue.*

291 "Von Neumann languages constantly keep": John Backus, "Can Programming Be Liberated from the von Neumann Style?" 1977 Turing Award Lecture, *Communications of the ACM,* August 1978, at http://portal.acm.org/affiliated/ citation.cfm?id=359579&dl=ACM&coll=ACM.

291 "When you learn about computer science": Jaron Lanier, quoted in Janice J.

Hess, "Coding from Scratch," Sun Developer Network, January 23, 2003, at http://java.sun.com/features/2003/01/lanier_qa1.html.

292 "Gordian software": Jaron Lanier, "Why Gordian Software Has Convinced Me to Believe in the Reality of Cats and Apples," Edge.org, November 19, 2003, at http://www.edge.org/3rd_culture/lanier03/lanier_index.html.

293 "If you make a small change": Lanier in Hess, "Coding from Scratch."

293 "The world as our nervous systems," "Try to be an ever better guesser," and "When you de-emphasize protocols": Lanier, "Gordian Software."

294 "very different and radical": Jaron Lanier talk at Future Salon, April 20, 2004. Information at http://www.futuresalon.org/2004/04/full_salon_with.html. Video at http://www.archive.org/movies/details-db.php?collection=open source_movies&collectionid=FutureSalon_04_2004.

294 "The moment programs grow beyond": Lanier, "Gordian Software."

294 "Little programs are so easy": Jaron Lanier talk at OOPSLA Conference, October 2004.

295 Daniel Dennett's critique of "Gordian Software" is at http://www.edge.org/discourse/gordian.html#dennett.

295 "The fundamental challenge for humanity": Jaron Lanier, interview with author, October 2005.

295 "I'm just sick of the stupidity": Jaron Lanier at OOPSLA 2004.

296 "We are stuck with the evolutionary pattern": Robert N. Britcher, The Limits of Software (Addison Wesley, 1999), p. 190.

296 "essential property" and following: Frederick Brooks, "No Silver Bullet: Essence and Accidents of Software Engineering," Computer 20:4 (April 1987), pp. 10–19.

297 "Computer science is in deep trouble": Gerald Jay Sussman, "Robust Design Through Diversity," September 11, 1999, at http://www.mindfully.org/GE/Robust-Design-Diversity-Sussman.htm.

297 David Lorge Parnas, "Software Aspects of Strategic Defense Systems," Communications of the ACM, December 1985, and at http://portal.acm.org/citation.cfm?id=214961&coll=ACM&dl=ACM.

299   "With code, the computer tells you": Bill Joy in a 2004 *New York Times* interview by Jon Gertner, June 6, 2004.

299   "My view is that we should": Richard Gabriel in Janice J. Hess, "The Poetry of Programming," Sun Developer Network, December 3, 2002, at http://java.sun.com/features/2002/11/gabriel_qa.html.

300   "It is as if all writers": From a slide set by Richard Gabriel called "Whither Software?" available at http://www.dreamsongs.com/Essays.html.

300   "I think we need to be ashamed" and "Everything we've done": Richard Gabriel talk at the Software Development Forum, Palo Alto, California, January 23, 2003.

302   "*art* meant something devised" and "The chief goal of my work": Donald Knuth, "Computer Programming as an Art," 1974 Turing Award lecture, in *Communications of the ACM*, December 1974.

303   "couldn't stand to write books": Donald Knuth quoted in Steve Ditlea, "Rewriting the Bible in 0's and 1's," *Technology Review*, September–October 1999.

304   "Beware of bugs in the above code": Knuth explains the exact origins of the much-cited quote at http://www-cs-faculty.stanford.edu/~knuth/faq.html.

304   "What were the lessons I learned": Donald Knuth, *Selected Papers on Computer Science* (CSLI Publicational/Cambridge University Press, 1996), p. 161.

305   "A longer attention span is needed": Ibid., p. 145.

306   The information about the Piet Hein poem over Knuth's entrance is from Ditlea, "Rewriting the Bible," in *Technology Review.*

306   "Instead of imagining that our main task": From Donald Knuth, "Literate Programming (1984)" in *Literate Programming*, Center for the Study of Language and Information, 1992, p. 99, as cited at http://www.literateprogramming.com/.

307   The "Linux kernel swear words" site is at http://www.vidarholen.net/contents/wordcount/.

308   The comments by the authors of the MyDoom, Bagle, and Netsky viruses are

reported in Iain Thomson, "Virus Writers Stage Online Slanging Match," Vnunet.com, March 3, 2004, at http://www.vnunet.com/vnunet/news/2124482/virus-writers-stage-online-slanging-match.

308 The Windows 2000 source code comments were reported in "We Are Morons: A Quick Look at the Win2k Source," Kuro5hin, February 16, 2004, at http://www.kuro5hin.org/story/2004/2/15/71552/7795.

CHAPTER 11
THE ROAD TO DOGFOOD

310 "To be effective at any large software": Jaron Lanier quoted in Janice Hess, "Coding from Scratch," Sun Developer Network, January 23, 2003, at http://java.sun.com/features/2003/01/lanier_qa1.html.

314 Ted Leung's blog posting looking for a "hot shot Python hacker," October 4, 2004, is at http://www.sauria.com/blog/2004/Oct/04.

314 "I was recently looking at the source": Phillip J. Eby, "Python Is Not Java," blog posting on December 2, 2004, at http://dirtsimple.org/2004/12/python-is-not-java.html.

317 "It is quite simply a thing of beauty": Phillip Eby blog posting from February 15, 2005, at http://dirtsimple.org/2005/02/making-it-from-scratch-with-tdd-and.html.

320 The line counts for Chandler were made by OSAF interns Brendan O'Connor and Arel Cordero.

320 The intern who compared joining Chandler to "moving to a new city" was Arel Cordero.

322 "While one cannot rule out the possibility": Ellen Spertus, "Why Are There So Few Female Computer Scientists?" MIT Artificial Intelligence Laboratory Technical Report 1315, August 1991, at http://people.mills.edu/spertus/Gender/pap/pap.html. A more recent update, "What We Can Learn from

Computer Science's Differences from other Sciences," is at http://www
.barnard.columbia.edu/bcrw/womenandwork/spertus.htm.

328 "This is the whole world of programming": Alan Kay's discussion of Lisp is in
"A Conversation with Alan Kay," *ACM Queue*, December 2004–January
2005, at http://acmqueue.com/modules.php?name=Content&pa=showpage&
pid=273.

328 Douglas Hofstadter, *Gödel, Escher, Bach: An Eternal Golden Braid* (Basic, 1979).

328 "Any recursive function will cause": From Colin Allen and Maneesh Dhagat,
LISP Primer, at http://grimpeur.tamu.edu/~colin/lp/node31.html.
My discussion of the halting problem is indebted to David Harel's lucid
explanations in his book *Computers Ltd.: What They Really Can't Do* (Ox-
ford, 2000).

329 "It is tempting to try and solve the problem": David Harel, *Computers Ltd.*,
p. 53.

330 "dashes our hope for a software system": Ibid., p. 50.

331 Hofstadter's Law appears on p. 152 of his book, *Gödel, Escher, Bach*.

333 "When people ask me when": Richard Stallman, quoted in Paul Jones,
"Brooks' Law and Open Source: The More the Merrier?" IBM Developer
Works, May 1, 2000, at http://www-128.ibm.com/developerworks/linux/
library/os-merrier.html.

334 Kapor's blog posting on the new power of Web-based software: "When
Browsers Grow Up," January 2, 2005, at http://blogs.osafoundation.org/
mitch/000812.html#000812.

335 Jon Udell's posting on the calendaring "train wreck" is at http://weblog
.infoworld.com/udell/2004/12/09.html#a1130.

335 Scott Mace's Calendar Swamp blog is at http://calendarswamp.blogspot.com/.

336 "I ran through a couple of crashers": Philippe Bossut's blog posting about
dogfooding Chandler, from October 3, 2005, is at http://wp.osafoundation
.org/2005/10/03/dogfooding-chandler/.

337 "Software development lacks one key element": Alan Cooper, *The Inmates
Are Running the Asylum* (SAMS, 1999), p. 41.

337 "Because the software is free": Linus Torvalds, quoted in Robert Cringely, "How Microsoft's Misunderstanding of Open Source Hurts Us All," October 23, 2003, at http://www.pbs.org/cringely/pulpit/pulpit20031023.html.

340 Firefox market share information is reported regularly at http://www.spread firefox.com/.

EPILOGUE
A LONG BET

347 The Bay Bridge delays were widely covered in the California press; for example, John King, "Towering Question: Will It Finally Be Built?" *San Francisco Chronicle*, March 21, 2006, at http://www.sfgate.com/cgi-bin/article .cgi?file=/c/a/2006/03/21/MNGD8HRJ761.DTL.

348 "the hard thing about building software": Frederick Brooks, "No Silver Bullet: Essence and Accidents of Software Engineering," *Computer* 20:4 (April 1987), pp. 10–19.

348 "The machines will become more and more": Freeman Dyson spoke at the O'Reilly Open Source Conference, Portland, Oregon, July 2004. Audio is available at http://www.itconversations.com/shows/detail170.html.

349 "Ultimately, information systems only give": Jaron Lanier at OOPSLA 2004 Conference.

349 Bill Gates in Singapore: Rohan Sullivan, "Gates Says Technology Will One Day Allow Computer Implants—but Hardwiring's Not for Him," Associated Press, July 5, 2005. Archived at http://www.interesting-people.org/archives/ interesting-people/200507/msg00029.html.

350 "relief from the confusions of the world": Ellen Ullman, *The Bug* (Nan A. Talese/Doubleday, 2003), p. 9.

352 Kapor's Long Bet with Ray Kurzweil is chronicled at http://www.longbets .org/1.

353 "radical transformation of the reality": Kurzweil described his vision of the Singularity in a talk hosted by the Long Now Foundation, San Francisco,

September 23, 2005. Video of the event is at http://video.google.com/video play?docid=610691660251309257.

353 "in the short term we always underestimate": Kurzweil, Long Now Foundation talk.

354 "As humans: We are embodied.": Kapor's essay accompanying the Long Bet is at http://www.longbets.org/1.

ACKNOWLEDGMENTS

This book could not have been written without Mitch Kapor's willingness to open the Chandler project's doors to me. At the time he first did so, neither of us expected that I would stick around as long as I did, or end up chronicling so many delays and setbacks. If he ever had second thoughts about exposing his team's work in this fashion, not once in the years I spent at OSAF did he bring them up. "It's the open source way," I know he would say.

For that I am deeply grateful, as I am to all the programmers and other staff at OSAF who welcomed me and took the time over the years to answer my questions about their work and themselves. The following people sat for extended interviews: John Anderson, Philippe Bossut, Donn Denman, Lisa Dusseault, Andy Hertzfeld, Mitch Kapor, Chao Lam, Ted Leung, Rys McCusker (via email), Lou Montulli, Sheila Mooney, Katie Parlante, Stuart Parmenter, Morgen Sagen, Brian Skinner, Michael Toy, Andi Vajda, and Mimi Yin. Many others shared their ideas and insights informally over lunch, in meeting rooms, or at their desks.

I'm grateful, too, to the legions of programmers and technologists who have chosen, over the past decade, to write publicly and candidly about their work online. Their work has made mine possible. Some of those whose writing and blogging I found especially helpful or relevant are Dan

Bricklin, Grady Booch, Adam Bosworth, Tim Bray, Geoff Cohen, Paul Ford, Martin Fowler, Paul Graham, David Heinemeyer Hansson, Robert Lefkowitz, Eric Sink, Joel Spolsky, Jon Udell, Dave Winer, and Jeremy Zawodny. Thanks to their work and that of many others, I'm convinced that future decades will look back on our time as a sort of golden age of Renaissance programmer-writers.

I also owe a debt of thanks to all my colleagues at Salon.com—in particular, to David Talbot, who gave me the opportunity to pursue this project, and to Joan Walsh, who welcomed me back once it was finished. Much of my thinking about programming has been informed by conversations and collaborations over the years with my friend and colleague Andrew Leonard. During the past decade, I've learned valuable lessons from everyone I've worked with in Salon's technology department, especially Chad Dickerson, who gave me early encouragement on this project.

Someday I hope to learn exactly how my agent, Stuart Krichevsky, wears his multiple hats—as advocate, diplomat, and adviser—so skillfully; his tireless efforts made mine easier at every stage. Thanks also to his staff, Shana Cohen and Elizabeth Kellermeyer. My editor at Crown, Rachel Klayman, embraced my ideas enthusiastically, championed them vigorously, and helped smooth their rough edges with a judicious and fiercely intelligent hand. I would also like to thank the rest of the team at Crown that helped launch this book into the world: Lucinda Bartley, Patricia Bozza, Tina Constable, Whitney Cookman, Maria Elias, Kristin Kiser, Donna Passannante, Philip Patrick, Steve Ross, Penny Simon, and Jie Yang.

Sue Halpern helped me sharpen my book proposal, for which much thanks. David Edelstein, Josh Kornbluth, and Bill McKibben read early versions of the book and provided much thoughtful feedback. Rafe Colburn and Mike Pence, programmers and writers both, read my manuscript and helped me make sure I wasn't driving into some technical ditch. Heikki Toivonen at OSAF also provided valuable comments and corrections.

Most authors thank their families; now I know why. My wife, Dayna

Macy, was the first reader for each of my chapters and guided my work toward clarity. Her encouragement and insights kept me on track each time the way forward seemed lost or blocked; her spirits kept mine aloft; her support and love made everything else possible.

My sons, Matthew and Jack, good-naturedly accepted this project's incursions into time they believed belong to them. They also shed light on the mysteries of the Lego Hypothesis.

My brother, Paul Rowe, set an example for me from childhood of eloquence and grace in written expression. (He also kindly put me up on my trips to New York.)

During my teens, my parents, Jeanne and Coleman Rosenberg—who doubtless would have preferred to see me pursue, say, cellular biology—gave me carte blanche to tinker with game programs, under the theory that "computers are the future." I think they're still right about that. This book is dedicated to them, with my gratitude for all their generosity and devotion, and my love.

INDEX

"free software," meaning of, 22

Fried, Jason, 260, 261, 263

Gabriel, Richard, 299–300
Game Neverending, 340
Gans, David, 35–36
Gates, Bill, 21, 26, 34, 36, 82, 169–70, 188, 203, 208, 349–50
geeks, 130–37
  behavioral profile of, 131–34
  evolution of word, 130–31
  informal communication mechanisms of, 135–37
"Geek Syndrome, The" (Silberman), 133–34
Gelernter, David, 166
generative programming, 278
Getting Things Done (Allen), 165–66, 337–38
Gillmor, Dan, 80, 81, 83, 85
Glass, Robert, 95
Gmail, 224, 260, 334
GNU, 21–22, 24
GNU-Linux, 22. See also Linux
Gödel, Escher, Bach (Hofstadter), 328
Google, 224, 260, 263, 266, 334, 340
"Gordian Software" (Lanier), 292–95
Gore, Al, 32, 33
Gosling, James, 143
GOTO statements, 239–40
government software:
  failed projects and,

47–49, 51–53, 247–48
  methodologies and, 244, 249, 254, 259
GPL (GNU Public License), 21
Graham, Paul, 79
groupware, 136–37
GTD (getting things done), 165–66, 337–38
GUI (graphical user interface), 47, 75, 101

halting problem, 329, 330, 337–38
Hansson, David Heinemeyer, 261, 262
Hardhat, 79
Harel, David, 329, 330
Harris, Jeffrey, 332
Hartsook, Pieter, 86, 233
Hawking, Stephen, 283–84
Hayes, Brian, 250
"Hello World" programs, 4–5
Hertzfeld, Andy, 11, 14, 30–31, 55, 60–61, 63, 85, 92–93, 104–5, 106, 109, 127, 150, 153, 159–60, 161, 163, 174, 187, 188, 207, 237, 263
  OSAF left by, 167–69
  selection of programming language and, 69–70, 71
  user interface and, 151, 152, 154–58
  Vista prototype and, 61, 70, 79, 151, 341
Hewlett-Packard, 128–29

Hofstadter, Douglas, 328, 331, 346
Holen, Vidar, 307–8
Hopper, Grace, 277
"Human Dynamics of Information Technology Teams, The," 132–33
Humphrey, Watts, 129–30, 241, 242–46, 247
Hungarian notation, 197–98

IBM, 25, 35, 51–52, 77, 86, 129, 177, 324
  System/360, 16–17, 242–43, 271
iCal, 209, 335, 336
IETF (Internet Engineering Task Force), 211–12
"Immaturity of CMM, The" (Bach), 245
infallibility, ascribed to computers, 351
infinite loops, 325–26, 328
Inmates Are Running the Asylum, The (Cooper), 337
In Search of Excellence (Peters), 128–29
integration, 160–61
Intentional Software, 278–80, 281, 283
Internet, 20, 24, 25, 38–39, 47, 96, 98, 211, 265–66, 299, 340
  standards for, 211–12
Internet Explorer, 66, 340
Internet time, 4
interpreted languages, 70–73. See also Perl; Python

ABOUT THE AUTHOR

Scott Rosenberg is cofounder of Salon.com, where he served as technology editor, then managing editor, and is now vice president for new projects. Before Salon he wrote on theater, movies, and technology for the *San Francisco Examiner* and was honored with the George Jean Nathan Award for his reviews. His writing has appeared in the *New York Times, Wired,* and many other publications. He lives in Berkeley, California, with his wife and two sons.

Visit his Web site at www.wordyard.com. This book's Web site is at www.dreamingincode.com.